Series on Chemistry, Energy and the Environment

Emerging Analytical Techniques for Chemical Speciation Studies

Part 1: Electrochemical and Migration Methods

Series on Chemistry, Energy and the Environment

ISSN: 2529-7716

The aim of this series on "Chemistry, Energy and the Environment" is to bring together authoritative contributions where the multidisciplinary character of Chemistry and its relationship to energy and the environment can be illustrated. In each volume, the latest advances in bio-, energy- and environmentally-related fields are described in a way that unambiguously shows the major contributions from chemists at the top of their field who describe new and improved technologies for the future.

Published

Vol. 15 *Emerging Analytical Techniques for Chemical Speciation Studies Part 1: Electrochemical and Migration Methods*
edited by Michel Meyer and Demetrio Milea

Vols. 11–14 *Synthesis and Applications in Chemistry and Materials (In 4 Volumes)*
Volume 11: Metal Coordination and Nanomaterials
Volume 12: Enzymatic and Organic Systems
Volume 13: Metal Complex Catalytic Systems and Materials
Volume 14: Biomass and Waste Valorisation, Functional Materials, Energy Conversion and Supercritical Systems
edited by Armando J. L. Pombeiro, Kamran T. Mahmudov and M. Fátima C. Guedes da Silva

Vol. 10 *Electrochemistry of Metalloporphyrins*
Karl M. Kadish, W. Ryan Osterloh and Eric Van Caemelbecke

Vol. 9 *Topics in Enantioselective Catalysis: Recent Achievements and Future Challenges*
edited by Angela Marinetti

For the complete list of volumes in this series, please visit
www.worldscientific.com/series/scee

Volume 15

Series on Chemistry, Energy and the Environment

Emerging Analytical Techniques for Chemical Speciation Studies

Part 1: Electrochemical and Migration Methods

Michel Meyer
Université de Bourgogne, France

Demetrio Milea
Università degli Studi di Messina, Italy

Series Editors

Karl M. Kadish
University of Houston, USA

Roger Guilard
Université de Bourgogne, France

World Scientific

NEW JERSEY • LONDON • SINGAPORE • BEIJING • SHANGHAI • HONG KONG • TAIPEI • CHENNAI • TOKYO

Published by

World Scientific Publishing Co. Pte. Ltd.
5 Toh Tuck Link, Singapore 596224
USA office: 27 Warren Street, Suite 401-402, Hackensack, NJ 07601
UK office: 57 Shelton Street, Covent Garden, London WC2H 9HE

British Library Cataloguing-in-Publication Data
A catalogue record for this book is available from the British Library.

Series on Chemistry, Energy and the Environment — Vol. 15
EMERGING ANALYTICAL TECHNIQUES FOR CHEMICAL SPECIATION STUDIES
Part 1: Electrochemical and Migration Methods

ISBN 978-981-12-7962-1 (hardcover)
ISBN 978-981-12-8109-9 (ebook for institutions)
ISBN 978-981-12-8110-5 (ebook for individuals)

For any available supplementary material, please visit
https://www.worldscientific.com/worldscibooks/10.1142/13537#t=suppl

Typeset by Stallion Press
Email: enquiries@stallionpress.com

https://doi.org/10.1142/9789811281099_fmatter

Preface

Thermodynamic studies of complexation equilibria (*e.g.* metal/ligand, host/guest, substrate/biological receptors) represent the core of many important branches of chemistry and biochemistry, while being the basis of chemical speciation analysis and modelling. Binding equilibria play a pivotal role in the design and synthesis of chemical compounds and drugs, their characterization and determination of their physical-chemical properties, as well as their implementation in specific applications. For example, complex formation or dissociation triggers the activity of many functional biomolecules and geochemical processes involved in precipitation or dissolution of minerals, sorption phenomena, or dissolution of gases in natural waters. Understanding the speciation of each constituent in a particular matrix or sample at a preset temperature and pressure, in other words knowing the distribution and concentration of all specific chemical forms in which the components might occur, is of broad interest and may lead to a wide range of applications. Acquiring complexation data (chemical composition and structural formulas, concentration-based stability constants, reaction enthalpies and entropies…) further allows for modelling and predicting the fate of metals and ligands in multicomponent systems under thermodynamic control, when varying, separately or simultaneously, one or more factors such as the concentration of components or competing agents, pH, temperature, pressure, salinity, ionic strength. Having access to these results then becomes an essential asset for workers and decision makers in fields as diverse as toxicology, clinical and biological chemistry, geochemistry, environmental and analytical chemistry, and industrial processes.

Chemical speciation and complexation thermodynamics are subbranches of analytical chemistry, as they rely on numerous instrumental analytical methods that provide directly, but most often indirectly, information on the concentration of relevant species in solution. Appropriate numerical processing of such data allows refining the sought after thermodynamic parameters, but also affords in some instances valuable additional electronic and/or structural information of the formed species. Historically, speciation analysis goes back about a century and some of the very earliest techniques are nowadays still routinely employed. The most popular of these methodologies is the so-called method of continuous variation, better known as the Job-plot in homage to the French chemist Paul Job, who devised in 1928 a simple UV spectrophotometric titration method for determining the stoichiometry of molecular complexes in solution. However, it is only in the early 1940s that Jannik Bjerrum pioneered the measurement of metal–ligand stability constants by employing a pH meter in conjunction with a glass electrode. With his seminal paper of 1941, Bjerrum developed a general method for the determination of stability constants relying on the change of pH in solution upon complex formation with ligands acting as weak bases. pH-potentiometry and spectrophotometry are typical examples of indirect approaches to unravel the sometimes quite intricate metal–ligand speciation, becoming over the years the two most popular methods.

Now, more than eighty years later, the field can today be judged as mature, as numerous other classical techniques have proven to be extremely powerful for evaluating binding affinities or association constants. When undertaking this project, it was not our purpose to review all of these techniques and thus produce a comprehensive treatise in the manner of the 18st-century encyclopedists. Some confidential or nowadays less-frequently used techniques have not been considered. We have also intentionally not included the most classical analytical methods, such as pH-potentiometry, spectrophotometry or calorimetry, as these have been treated in many reviews, tutorials, and books.

While studies of solution equilibrium chemistry might be described as flourishing from the 1960s until the turn of last century, it has remained the apanage of specialists during that period. Since then, the advent of modern computers, powerful numerical tools, and user-friendly low cost instrumentation opened up the field to researchers in multiple disciplines who could easily and quickly collect the data despite lacking in many cases a full understanding of the solution equilibria concepts. Because of this, we are today facing a paradox, as modern instrumentation and techniques have never been so powerful and progressing so rapidly, while in-depth knowledge of the theory is becoming rarer with the turn of generations.

Although several monographs focusing on a specific analytical technique described in this volume have appeared in the past, there is no recently published general textbook or treatise dealing specifically with speciation studies and the measurement of thermodynamic parameters by the wide panel of techniques currently available to users in the scientific community. We therefore felt that the time was ripe to collect a series of chapters in the form of a guide or Handbook, which describes both classical and emerging techniques, covers theoretical and practical aspects, and at the same time provides helpful experimental guidelines to the end user. As the first of a series of two volumes, this book assembles four chapters focused on electrochemical and migration techniques. The second volume, still in preparation, will cover spectroscopic methods.

Each chapter has been authored by eminent practitioners, highlighting the latest technical and computational advances in a way that unambiguously show the major contributions of the authors who are at the top of their field. Hence, these volumes are intended to be a valuable introduction to both students and senior scientists not in the field, while at the same time being a helpful companion to more experienced users of each instrumental technique. The chapters provide an up-to-date overview with useful tips and hints on the application of selected cutting-edge analytical methods that allow one to unravel and model intricate complex formation equilibria. In summary, these volumes should be of interest to a broad readership, including researchers in the field of analytical, coordination, supramolecular, environmental, instrumental, and physical chemistry as well as radiochemists, electrochemists, and biochemists.

We would like to close by expressing our gratitude to the contributing authors, both for their chapters and for their constructive exchanges. We hope that their efforts in bringing together this collection will be valuable to many scientists in the future while also fostering new directions of research. However, when looking to the future, one cannot forget the past. We are in debt to our predecessors who have paved the way forward, leading to many current state-of-the-art studies in the field of chemical speciation. We would also like to dedicate this volume and the one that follows to the memory of our respective mentors who sadly passed away after the start of the project but before its completion, namely Dr. Anne-Marie Albrecht-Gary (Strasbourg) and Prof. Silvio Sammartano (Messina), who introduced us to solution equilibrium studies. In addition, we would like to thank Professors Kenneth N. Raymond (Berkeley), Klaus Bernauer (Neuchâtel), Roger Guilard (Dijon), Concetta De Stefano (Messina), and Frank J. Millero (Miami), for their guidance at the early stage of our career. Being their co-workers for some time has been a unique training experience and a privilege for us.

A special acknowledgement is also given to all of our friends and colleagues in the International Group for the Thermodynamics of Complexes (the ISMEC Group) for giving us, over many years, the opportunity to find the right environment for a fruitful and more comprehensive understanding of these topics. We also want to thank Dr. Sofia Gama (Lisbon) for her dedication, competence, and extraordinary patience in working with both of us.

Finally, we wish to dedicate this book to our families and, especially, DM to his beloved daughter Benedetta and son Domenico, for "paying the price" of our dedication to Science.

June 2024

Michel Meyer, Dijon
Demetrio Milea, Messina

https://doi.org/10.1142/9789811281099_fmatter

Contents

https://doi.org/10.1142/9789811281099_0001

1 A Unified Approach in the Voltammetric and Potentiometric Studies of Metal–Ligand Equilibria: Theory and Practical Guide

Ignacy Cukrowski,[*,†] Caren Billing[‡] and Tumaini Mkwizu[‡]

[†]Department of Chemistry, Faculty of Natural and Agricultural Sciences, University of Pretoria, Hatfield, Lynnwood Road, Pretoria, 0002, South Africa
[‡]School of Chemistry, Faculty of Science, University of the Witwatersrand, 1 Jan Smuts Avenue, Braamfontein, Johannesburg, 2000, South Africa
*ignacy.cukrowski@up.ac.za

Table of Contents

I. List of Abbreviations

ADC	analog-to-digital converter
AE	amalgam electrode
CCFC	computed complex formation curve
CFC	complex formation curve, either experimental (ECFC) or computed (CCFC)
CS	corrected shift of polarographic signal
D	diffusion coefficient
DAC	digital-to-analog converter
DAQ	data acquisition
DCP	direct current polarography
DME	dropping mercury electrode
DPP	differential pulse polarography
ECFC	experimental complex formation curve
ECFC-pH	experimental complex formation curve plotted *vs*. pH of a sample solution

ECFC-pL	experimental complex formation curve plotted *vs.* pL = –log[L]
ESTA	Equilibrium Simulation by Titration Analysis: software dedicated to potentiometry
GE	glass electrode
GEP	glass electrode potentiometry
HEDP	1-hydroxyethylenediphosphonic acid
ISE	ion selective electrode
MBEs	mass-balance equations
ME	metallic electrode
MLE	metal–ligand equilibrium
NNS	new Nernstian slope
NNS-pH	new Nernstian slope obtained by plotting CS *vs.* pH
NNS-pL	new Nernstian slope obtained by plotting CS *vs.* pL
SDD	species distribution diagram
Tpen	*N*,*N*,*N′*,*N′*-tetra-2-picolylethylenediamine

II. Introduction

The first report on polarographic studies of labile complexes was published in 1941 by Lingane.[1] This method was very limited as it allowed studying a ligand–metal system containing just a single (ML or ML_2) metal species. Ten years later (in 1951), DeFord and Hume[2] improved on Lingane's protocol and showed that one can investigate ML_q complexes formed in a stepwise fashion. The formation constants, however, had to differ such that each complex dominated other species significantly under experimental conditions employed. These methods allowed studying simple ligands forming moderate (weak) metal complexes.[3] Schaap and McMasters[4] proposed an experimentally demanding and somewhat tedious[5] modification of the straightforward DeFord-Hume approach; nonetheless, two simple ligands competing for a single metal ion could be studied when ternary $[M(L1)_j(L2)_k]$complexes were formed. The focus of all the above methods was solely on labile metal-ligand systems studied at fixed pH; the shift in the half-wave potential was used for the evaluation of stability constants. In each case, formation of complexes was achieved by increasing an excess of a ligand.

Actually, the excess of a ligand had to be large, otherwise the calculated formation constants might be unreliable.[6] Exactly the same experimental conditions, *i.e.*, ligand titrations performed on a sample with fixed pH, were employed in highly specialized approaches reported for polarographic studies of (i) non-labile[7,8] metal complexes, with chelating ligands where the variation in signal intensity was utilized in determining formation constants when two metal ions competed for a single ligand, or (ii) polyelectrolytes.[9,10]

Most reported formation and protonation constants[11,12] come from glass electrode potentiometry (GEP).[13] GEP is a simple and extremely convenient tool that allows for studying complexes formed between a countless number of organic ligands and the majority of metal ions in a wide pH-range. However, as with any analytical method, it has a few severe limitations, among them:

1. It does not work well, if at all, at very low and high pH values. Hence, studying highly acidic metal ions, such as Bi(III), or complexes formed at pH > 12, is a challenge.
2. It requires comparable and relatively high total concentrations of the metal ion and ligand. This means that often hydrolysis followed by precipitation of metal containing species cannot be easily avoided. In addition, formation of weak metal complexes cannot be "promoted" by a large increase in the total ligand concentration.

Clearly, the above limitations of GEP open up a window of opportunity for other analytical techniques. Interestingly, polarography has never become a powerful and reliable method in the field even though it was employed in the study of metal complexes before GEP and it can, in principle, be used where GEP fails. This was demonstrated when differential pulse polarography (DPP) was used in the field of ligand design strategies. It provided reliable formation constants for both labile and non-labile complexes[14–17] when experiments were performed using classical GEP conditions, namely, at the fixed ligand-to-metal ratio and varied pH. Due to the intrinsic properties of polarography, strong complexes of very acidic Bi(III)[14,15] (formed at low pH) and relatively weak complexes of Pb(II)[16,17] with organic multi-dentate ligands were investigated. Often a low total metal ion concentration and a large ligand-to-metal ratio were used to suppress metal ion hydrolysis when pH was varied;[14–18] the condition that cannot be met when GEP is employed. In order to study protonated and hydroxo complexes at a fixed total ligand $[L]_T$ to total metal ion $[M]_T$ concentration ratio, the classic Lingane equation was modified.[15–17]

A real break through took place in 1996 when a new method of speciation was published.[19] A single mathematical expression was derived that was applicable to any kind of $M_pL_qH_r$ complexes investigated by acid–base titration, as it is done in GEP experiments. Also and for the first time in polarographic studies, this new expression required making use of mass-balance equations (MBEs) to compute the free metal ion and free ligand concentrations needed to compute formation constants. Complex formation curves were introduced to the field and proved to be a formidable tool in modelling solution equilibria and refinement of formation constants.[20] The power of the new school of speciation by polarography was demonstrated on many metal–ligand systems[21–46] and this included the first report on polynuclear species determined by polarography. Reliability of M–L–OH models and constants computed were, wherever possible, verified either from separate DPP, direct current polarographic (DCP) and GEP experiments or even from joint DPP/GEP or DCP/GEP experiments. It has also been shown that the general expression reported in 1996[19] and related complex formation curves (experimental and computed introduced the next year[20]) are also applicable to potentiometric studies by use of ion selective electrodes (ISEs) and running combined polarographic and ISE experiments on a single sample solution became a reality. Finally, polarographic data were transformed to virtual potentiometric format suitable for refinement operations in software dedicated to potentiometry, such as ESTA.[47–50] Actually, the new school of speciation can be seen as a united multi-analytical approach that provides a complete and unique platform; it is suitable to metal complexes in wide ranges of the total ligand and metal ion concentrations.

This chapter is written as a dedicated tutorial to guide the reader and potential user in designing and executing experiments and interpreting recorded and simulated data. The speciation study (by any analytical technique) is a multi-stage process; each one is explained and data from real experiments were used for illustration purposes. It is our hope that content of this chapter will be found useful even for non-electrochemists.

III. Nomenclature Used and Applicable Mathematical Models

A. Nomenclature

Basic nomenclature and expressions for formation constants that are generally applicable in the field of metal–ligand equilibria studies, and employed here, are introduced in this section. For simplicity, charges on all species are omitted.

The main focus is on equilibria involving ligands that undergo protonation reactions in water. Hence, a metal ion M might react with either a free ligand L or some protonated form H_nL. In such a case, we can write a general expression for a complex formation reaction as:

$$p\text{M} + q\text{L} + r\text{H} \rightleftarrows \text{M}_p\text{L}_q\text{H}_r \tag{1}$$

for which the overall concentration formation constant is defined as:

$$\beta_{pqr} = \beta_{\text{M}_p\text{L}_q\text{H}_r} = \frac{[\text{M}_p\text{L}_q\text{H}_r]}{[\text{M}]^p[\text{L}]^q[\text{H}]^r} \tag{2}$$

where molar concentrations of all species are used rather than activities. Most practical applications require complexation data at a specific ionic strength (I) of a solution sample, the presence of a background electrolyte at a specific molar concentration (*e.g.*, 0.13 M NaCl solution is used in modelling blood plasma) and a specific temperature (*e.g.*, 37 °C is suitable for blood plasma modelling). The ionic strength and temperature are kept fixed throughout an experiment when formation constants are determined in selected background electrolyte. The thermodynamic overall formation constant β_{pqr}^{th} can be computed when activity coefficients γ of all species are known (Eq. 3).

$$\beta_{pqr}^{\text{th}} = \frac{a(\text{M}_p\text{L}_q\text{H}_r)}{(a_\text{M})^p(a_\text{L})^q(a_\text{H})^r} = \frac{[\text{M}_p\text{L}_q\text{H}_r]}{[\text{M}]^p[\text{L}]^q[\text{H}]^r} \times \frac{\gamma(\text{M}_p\text{L}_q\text{H}_r)}{(\gamma_\text{M})^p(\gamma_\text{L})^q(\gamma_\text{H})^r}$$
$$= \beta_{pqr} \times \frac{\gamma(\text{M}_p\text{L}_q\text{H}_r)}{(\gamma_\text{M})^p(\gamma_\text{L})^q(\gamma_\text{H})^r} \tag{3}$$

Complexes are formed in a stepwise manner, *e.g.*, $\text{M} + \text{L} \rightleftarrows \text{ML}$, $\text{ML} + \text{L} \rightleftarrows \text{ML}_2$, $\text{ML}_2 + \text{L} \rightleftarrows \text{ML}_3$, *etc*. Moreover, M, L and all complexes exist in solution at any time and their %-fraction varies with pH of a solution, temperature, and the kind of background electrolyte. Equilibrium for each consecutive complex formation step (1, 2, 3, …, n) is described by the stepwise concentration equilibrium constants:

$$\mathrm{M + L \rightleftarrows ML}; \quad K_1 = K_{\mathrm{ML}} = \frac{[\mathrm{ML}]}{[\mathrm{M}][\mathrm{L}]} \tag{4}$$

$$\mathrm{ML + L \rightleftarrows ML_2}; \quad K_2 = K_{\mathrm{ML_2}} = \frac{[\mathrm{ML_2}]}{[\mathrm{ML}][\mathrm{L}]} \tag{5}$$

$$\mathrm{ML_2 + L \rightleftarrows ML_3}; \quad K_3 = K_{\mathrm{ML_3}} = \frac{[\mathrm{ML_3}]}{[\mathrm{ML_2}][\mathrm{L}]} \tag{6}$$

Having stepwise formation constants, one can calculate the overall constants from:

$$\beta_1 = \beta_{110} = K_1 \quad \log \beta_1 = \log K_1 \tag{7}$$

$$\beta_2 = \beta_{120} = K_1 \times K_2 \quad \log \beta_2 = \log K_1 + \log K_2 \tag{8}$$

$$\beta_3 = \beta_{130} = K_1 \times K_2 \times K_3 \quad \log \beta_3 = \log K_1 + \log K_2 + \log K_3 \tag{9}$$

Many organic ligands contain –COOH groups from which a proton can dissociate and/or atoms, such as N, that can be protonated. For simplicity and convenience, all reactions between a ligand and a proton H are considered as protonation processes. Hence, the free form of a ligand L is the one that does not have a proton that can dissociate. Equilibria and expressions for equilibrium constants can be written for the purpose of ligand protonation reactions and stepwise concentration protonation constants:

$$\mathrm{H + L \rightleftarrows HL}; \quad K_1^{\mathrm{H}} = \frac{[\mathrm{HL}]}{[\mathrm{H}][\mathrm{L}]} \tag{10}$$

$$\mathrm{HL + H \rightleftarrows H_2L}; \quad K_2^{\mathrm{H}} = \frac{[\mathrm{H_2L}]}{[\mathrm{HL}][\mathrm{H}]} \tag{11}$$

$$\mathrm{H_2L + H \rightleftarrows H_3L}; \quad K_3^{\mathrm{H}} = \frac{[\mathrm{H_3L}]}{[\mathrm{H_2L}][\mathrm{H}]} \tag{12}$$

Relationships described by Eqs. 7–9 also hold for the stepwise protonation reactions and, as an example, for the second protonation the following holds:

$$\beta_2^{\mathrm{H}} = \beta_{012}^{\mathrm{H}} = K_1^{\mathrm{H}} \times K_2^{\mathrm{H}} \quad \log \beta_2^{\mathrm{H}} = \log K_1^{\mathrm{H}} + \log K_2^{\mathrm{H}} \tag{13}$$

Realizing that the last (*n*th) dissociation step can be also seen as the first protonation reaction, *i.e.*, the dissociation reaction HL ⇄ H + L can be written as a protonation reaction H + L ⇄ HL, there is a simple relationship between dissociation $K_{\mathrm{a}_n}^{\mathrm{H}}$ and protonation K_n^{H} constants. Assuming that a ligand has three dissociable protons, then:

$$K_{\mathrm{a}_1}^{\mathrm{H}} = \frac{1}{K_3^{\mathrm{H}}} \quad K_{\mathrm{a}_2}^{\mathrm{H}} = \frac{1}{K_2^{\mathrm{H}}} \quad K_{\mathrm{a}_3}^{\mathrm{H}} = \frac{1}{K_1^{\mathrm{H}}} \tag{14}$$

Typically, the formation of complexes with protonated ligands, *e.g.*, MLH or M(HL), are expected at lower pH and they always form prior to the formation of complexes with a free ligand, like ML or ML_2. Furthermore, polynuclear complexes would be expected to start forming prior to mononuclear ones. Hence, M_2L is expected to form before ML which, in turn, might be followed by ML_2 formation. At specific pH values some metal ions and their complexes might undergo hydrolysis. One must realize that due to constancy in the K_{w} = [H][OH] value at a selected temperature and ionic strength of a solution (*e.g.* $K_{\mathrm{w}} = 10^{-13.74}$ at 25 °C and I = 0.5 M (H,Na)NO_3), the OH^- anions are always available, regardless of the pH of a solution. In fact, some M ions and their complexes might undergo hydrolysis at relatively low pH values (or in highly acidic media); clearly, coordination of OH groups must also be accounted for. For hydrolyzed complexes we can write:

$$p\mathrm{M} + q\mathrm{L} + r\mathrm{OH} \rightleftarrows \mathrm{M}_p\mathrm{L}_q(\mathrm{OH})_r \tag{15}$$

for which the overall concentration formation constant is:

$$\beta_{pq-r} = \beta_{\mathrm{M}_p\mathrm{L}_q(\mathrm{OH})_r} = \frac{[\mathrm{M}_p\mathrm{L}_q(\mathrm{OH})_r]}{[\mathrm{M}]^p[\mathrm{L}]^q[\mathrm{OH}]^r} \tag{16}$$

and, as an example, $\beta_{12-1} = [ML_2(OH)]/[M][L]^2[OH]$ or $\beta_{11-2} = [ML(OH)_2]/[M][L][OH]^2$ represent the overall concentration formation constant of $ML_2(OH)$ and $ML(OH)_2$, respectively.

B. Mathematical Expressions Applicable to Speciation by Voltammetry

Strictly speaking, expressions included in this section apply to (i) homogeneous reactions in a solution that equilibrate fast (labile complexes, protonation reactions) and (ii) an electron transfer which can be considered as a reversible electrochemical process, *i.e.*, with fast heterogeneous kinetics. For convenience and simplicity, let us consider all processes taking place at the electrode-solution interface as made mainly of two reactions (or processes) — see Scheme 1.

The first (controlled by homogenous kinetics) involves dynamic equilibria between L, M and their complexes. The second process (controlled by heterogeneous kinetics) is the actual electrochemical reaction of the electroactive species, which may be a metal complex of lower coordination number than ML_q or the aqua metal ion. Several critical factors have to be considered when studying metal–ligand systems by polarography, namely:

1. The presence of side reactions.
2. The association and dissociation rates of the complexes relative to the timescales of the measurements; this makes the complexes labile, non-labile or quasi-labile on the selected time-scale.
3. Electron-transfer kinetics at the electrode surface relative to the timescale of the measurements; this determines the electrochemical processes as reversible, quasi-reversible or irreversible.

homogenous kinetics

$$\underbrace{ML_q(aq) \rightleftarrows ML_{q-1}(aq) + L(aq) \rightleftarrows \ldots \rightleftarrows M(aq) + qL(aq)}$$

$$\uparrow\downarrow \; + ne,\, k_s \text{ heterogeneous kinetics}$$

$$M(Hg) + xL, \text{ where } 0 \le x \le q$$

Scheme 1. Schematic and highly simplified presentation of processes occurring at the solution-electrode interface during the reduction of metal complexes.

The methodology discussed in detail in the sections that follow is applicable to numerous scenarios, among them:

1. Diverse titration methods: (a) measurements at fixed $[L]_T$:$[M]_T$ and various pH values, (b) titration with a ligand solution (typically at a fixed pH) and (c) titration with a metal ion solution. In principle, any combination of all the above can also be considered.
2. Wide-ranging types of complexes: consecutively formed $M_pL_qH_r$ (including polynuclear metal species) or $M_pL_q(OH)_r$ complexes.
3. Kinetically complicated metal–ligand systems: fully labile, fully inert, intermediate, and kinetically mixed metal–ligand systems.
4. Diverse electrochemical techniques: polarography (*e.g.*, differential pulse and direct current modes, DPP and DCP, respectively) as well as free metal ion sensitive probe, such as ion selective electrode (ISE), metallic and amalgam electrodes (ME and AE, respectively).

In the following sections notation typically pertaining to DPP studies will be used, but all the mathematical expressions are equally applicable to the DCP studies. The following equation, reported in 1996,[19] is directly used in polarographic studies:

$$E_{\mathrm{p}}^{\mathrm{r}}(\mathrm{M}) - E_{\mathrm{p}}^{\mathrm{r}}(\mathrm{comp})_{\mathrm{x}(i)} - \frac{RT}{nF}\ln\left(\frac{I_{\mathrm{p}}^{\mathrm{r}}(\mathrm{comp})}{I_{\mathrm{p}}^{\mathrm{r}}(\mathrm{M})}\right)_{\mathrm{x}(i)} = \frac{RT}{nF}\ln\left(\frac{[\mathrm{M}]_{\mathrm{T}}}{[\mathrm{M}]}\right)_{\mathrm{x}(i)} \tag{17}$$

In Eq. 17, *R*, *T*, *n* and *F* have the same meaning as in the classical Nernst equation. It is important to stress that Eq. 17 is only valid when relevant data represent an electrochemically reversible process. Hence, the superscript "r" features in the notation used for DPP peak potentials ($E_{\mathrm{p}}^{\mathrm{r}}$) and currents ($I_{\mathrm{p}}^{\mathrm{r}}$) in Eq. 17. To this effect, protocols useful in testing electrochemical reversibility are described in Section XII. The $E_{\mathrm{p}}^{\mathrm{r}}(\mathrm{M})$ term stands for the reduction DPP peak potential recorded on initial volume, *V*(init), of a free metal ion solution. In other words, $E_{\mathrm{p}}^{\mathrm{r}}(\mathrm{M})$ represents a placement of a polarographic signal on the scale of potentials applied to a working electrode when a curve was recorded in the presence of a metal ion but in the absence of a ligand (the same applies to $E_{1/2}^{\mathrm{r}}(\mathrm{M})$ when DCP is used). All remaining terms have a subscript x(*i*) that stands for either pH(i) when measurements are taken during titrations at a fixed $[L]_T$:$[M]_T$ ratio and varying pH values, or pL(i) when solution is kept at a fixed pH and the total ligand concentration is varied. The $E_p^r(\text{comp})_{x(i)}$ term has the same meaning as $E_p^r(\text{M})$ except it refers to the reduction potential obtained from a DPP peak recorded in the presence of a ligand in the solution sample when complexes are expected to form. The terms $I_p^r(\text{M})_{x(i)}$ and $I_p^r(\text{comp})_{x(i)}$ stand for the intensities of the reversible polarographic signals (DPP peaks in this case) obtained at a specific pH(i) or pL(i) value (in the case of DCP, limiting diffusion currents $I_d^r(\text{M})_{x(i)}$ and $I_d^r(\text{comp})_{x(i)}$ apply). Importantly, whereas the $I_p^r(\text{comp})_{x(i)}$ term comes directly from experimental data recorded at x(i) in the presence of a ligand, the $I_p^r(\text{M})_{x(i)}$ term is not a measurable value as it represents an expected intensity one should observe for a free metal ion when no complexes are formed. Fortunately, the expected intensity of a free metal ion can be calculated by accounting for dilution of the sample solution at each x(i), *i.e.*, $I_p^r(\text{M})_{x(i)} = I_p^r(\text{M}) \times V(\text{init})/V_{x(i)}$ where $I_p^r(\text{M})$ was obtained for the initial sample without a ligand present (notably, it does not feature in Eq. 17) and $V_{x(i)} = V(\text{init}) + V(\text{titrant})$ is the total volume of a sample solution at x(i) after the addition of V(titrant) to the initial volume of a sample solution V(init). From this follows that $I_p^r(\text{M})_{x(i)} < I_p^r(\text{M})$ holds at all x(i) values as the dilution factor $V(\text{init})/V_{x(i)}$ is always a fraction. The $([M]_T)_{x(i)}$ term stands for the total metal ion concentration at each experimental point and, due to dilution of the sample solution throughout a titration, it is always smaller than the free metal ion concentration before addition of a ligand, *i.e.*, the very first sample solution with $[M]_T$ for which $E_p^r(\text{M})$ and $I_p^r(\text{M})$ were recorded. Finally, the $[M]_{x(i)}$ represents the free metal ion concentration in the presence of ligand at every x(i) value, *i.e.* the concentration of the metal ion that remains uncomplexed as the pH or ligand concentration is changed. Clearly, if metal complexes are formed then the $[M]_{x(i)}$ value is unavailable from such an experiment. The only way to obtain this value is by solving appropriate mass-balance equations (MBEs) for the assumed metal–ligand model.

Since $I_p^r(\text{M})_{x(i)}$ and $([M]_T)_{x(i)}$ must be calculated at each x(i), the added volume of a titrant at each consecutive step, $V(\text{titrant})_{x(i)}$, must be recorded as accurately as possible, *e.g.*, to a small fraction of a mL (when a burette is used) or a μL when a micro syringe is used.

The left-hand side of Eq. 17 is calculated from experimental data and is called the corrected shift (CS) of a polarographic signal when complexes are formed. If an experiment is conducted at a fixed total ligand ($[L]_T$) and total metal ion ($[M]_T$) concentration ratio ($[L]_T$:$[M]_T$ = constant) and the pH of the sample solution is varied, then the corrected shift is computed for each recorded polarogram at a specific pH(i) value according to:

$$CS_{\text{pH}(i)} = E_{\text{p}}^{\text{r}}(\text{M}) - E_{\text{p}}^{\text{r}}(\text{comp})_{\text{pH}(i)} - \frac{RT}{nF}\ln\left(\frac{I_{\text{p}}^{\text{r}}(\text{comp})}{I_{\text{p}}^{\text{r}}(\text{M})}\right)_{\text{pH}(i)} \tag{18}$$

Typically, when complexes are formed in solution, the recorded polarogram shifts towards more negative applied potentials and its intensity decreases. This means that the following should hold throughout an experiment involving an acid–base titration:

1. $\Delta E_{\text{pH}(i)}^{\text{r}} = E^{\text{r}}(\text{M}) - E^{\text{r}}(\text{comp})_{\text{pH}(i)} > 0$,
2. $\{I^{\text{r}}(\text{comp})/I^{\text{r}}(\text{M})\}_{\text{pH}(i)} < 1$ hence $\ln\{I^{\text{r}}(\text{comp})/I^{\text{r}}(\text{M})\}_{\text{pH}(i)} < 0$.

Consequently, the corrected shift $CS_{\text{pH}(i)}$ is always positive and $CS_{\text{pH}(i)} > \Delta E_{\text{pH}(i)}^{\text{r}}$ applies as long as $I^{\text{r}}(\text{comp}) < I^{\text{r}}(\text{M})$ holds. In other words, the corrected shift predicts by how much the signal should have shifted if its intensity, after accounting for dilution due to titrant added, remained the same throughout an experiment. The significance of this will be explained in Section VIII.

To fully understand and appreciate what it takes to rigorously compute [M], the only term in Eq. 17 that is unavailable from a given experiment, let us consider the very simple and well-studied Cd–glycine–OH system. Besides the free metal ion, the following known complexes with glycine (M(HL), ML, ML_2, ML_3, $ML_2(OH)$) and all known hydroxo complexes of cadmium (M(OH), $M(OH)_2$, $M(OH)_3$, $M(OH)_4$, $M_2(OH)$ and $M_4(OH)_4$) must be incorporated in the expression for the total metal ion concentration:

$$\begin{aligned}[\text{M}]_{\text{T}} = {} & [\text{M}] + [\text{M(HL)}] + [\text{ML}] + [\text{ML}_2] + [\text{ML}_3] \\ & + [\text{ML}_2(\text{OH})] + [\text{M(OH)}] + [\text{M(OH)}_2] \\ & + [\text{M(OH)}_3] + [\text{M(OH)}_4] + 2[\text{M}_2(\text{OH})] + 4[\text{M}_4(\text{OH})_4]\end{aligned} \tag{19}$$

Except $[M]_T$, none of the individual species concentration terms in Eq. 19 is known from an experiment — there are 12 unknowns. However, one can express each concentration term in Eq. 19 using the relevant overall concentration formation constant. This results in the expression:

$$\begin{aligned}[M]_T = {} & [M] + \beta_{M(HL)}[M][H][L] + \beta_{ML}[M][L] + \beta_{ML_2}[M][L]^2 \\ & + \beta_{ML_3}[M][L]^3 + \beta_{ML_2(OH)}[M][L]^2[OH] \\ & + \beta_{M(OH)}[M][OH] + \beta_{M(OH)_2}[M][OH]^2 \\ & + \beta_{M(OH)_3}[M][OH]^3 + \beta_{M(OH)_4}[M][OH]^4 \\ & + 2\beta_{M_2(OH)}[M]^2[OH] + 4\beta_{M_4(OH)_4}[M]^4[OH]^4\end{aligned} \tag{20}$$

where the total number of unknown concentration terms has been reduced to two, namely the free metal ion concentration [M] and the free ligand concentration [L]. It is important to note that the free proton concentration [H] is obtained from pH-measurements and [OH] is computed from the known water dissociation constant K_w and known [H]. It is assumed that all data pertaining to $M_p(OH)_r$ species are known and must be incorporated in the M–L–OH model. It is obvious that the number of β values to be determined from a polarographic experiment depends on the assumed M–L–OH model. Protocols that are useful in building up the most likely M–L–OH models and estimates in log β values for M and L containing species, all based on experimental data collected, are described in the subsequent sections.

To compute the total ligand concentration one must sum up concentrations of all ligand containing species and in this particular case, the following applies:

$$\begin{aligned}[L]_T = {} & [L] + [HL] + [H_2L] + [M(HL)] + [ML] + 2[ML_2] \\ & + 3[ML_3] + 2[ML_2(OH)]\end{aligned} \tag{21}$$

As before, the concentration terms can be expressed using the relevant overall protonation/formation constants and this results in:

$$\begin{aligned}[L]_T = {} & [L] + \beta_1^H[H][L] + \beta_2^H[H]^2[L] + \beta_{M(HL)}[M][H][L] \\ & + \beta_{ML}[M][L] + 2\beta_{ML_2}[M][L]^2 \\ & + 3\beta_{ML_3}[M][L]^3 + 2\beta_{ML_2(OH)}[M][L]^2[OH]\end{aligned} \tag{22}$$

It is important to realize that, due to analytical procedures commonly applied in the study of metal-ligand equilibria, the total molar concentrations of metal $[M]_T$ and ligand $[L]_T$ as well as the free proton concentration [H] are known at each titration step. To minimize a propagation of errors, these concentrations should be expressed to four significant figures after a decimal point as, *e.g.*, $[Cd^{2+}]_T = 2.5397 \times 10^{-4}$ M or [H] = 0.0959 M. Also, all protonation constants for a ligand must be known. Hence, there are only two unknown concentration terms common to Eqs. 20 and 22, *i.e.*, [M] and [L]. These two concentration terms can only be obtained by solving MBEs for the assumed metal–ligand system. It is important to recall (as clearly stated already in Section III.A) that to determine the reliable overall (or stepwise) concentration formation constants one must assure constancy in the ionic strength I of a sample solution and solution temperature throughout the entire experiment. This means that a background electrolyte used should be about 100 times more concentrated than the total ligand concentration that, in turn, should be at least 30 (preferably 50) times more concentrated than the total metal ion concentration. The indicated ratios also depend on the number of protonation constants of a ligand and the pH-range within which data is collected as this impacts the number and concentration of charged ligand species, as well as degree of lability of metal complexes formed.

It is known that to solve expressions with k-unknowns, the minimum number of required equations is also k. It is also known that to obtain the best quality formation constants (our unknowns), with the lowest standard deviations, one should solve many more equations. To achieve that, one must record significantly more polarograms, preferably 2–3 times more than the number of unknowns. Note that each polarogram provides experimental data for $CS_{\text{pH}(i)}$ (Eq. 18) and a pair of MBEs (Eqs. 20 and 22), all applicable to a specific pH(i) value (or x(i) value in general).

C. Mathematical Expressions Applicable to Speciation by Potentiometry

It is generally accepted that the response of an ISE is toward the thermodynamic activity of a free metal ion. However, working with a large enough excess of a background electrolyte and maintaining I and T

constant assures a linear response of an ISE toward the free metal ion concentration due to the activity coefficients remaining essentially constant. Ideally, such a response should not be negatively affected by the formation of metal complexes. This means that a perfect ISE should generate a linear response with a near Nernstian slope in a limited concentration range of M. To assess the possible use of Eq. 17, we note that:

1. Only potentials are measured with an ISE. As one does not record current when an ISE is used, the current-containing term in Eq. 18 is not applicable and the corrected shift is reduced to $\Delta E^{\mathrm{r}}_{\mathrm{pH}(i)} = E^{\mathrm{r}}_{\mathrm{p}}(\mathrm{M}) - E^{\mathrm{r}}_{\mathrm{p}}(\mathrm{comp})_{\mathrm{pH}(i)}$. Hence, the corrected shift is simply a change in a potential of an ISE, $\Delta E_{\mathrm{ISE,pH}(i)}$, calculated at each pH($i$) value relative to the potential recorded for a solution sample containing the initial metal ion concentration [M] without ligand added, *i.e.*, $\Delta E_{\mathrm{ISE,pH}(i)} = E_{\mathrm{ISE}}(\mathrm{M}) - E_{\mathrm{ISE}}(\mathrm{comp})_{\mathrm{pH}(i)}$.
2. The right-hand side of Eq. 17 remains unchanged.

To this effect, for the initial solution sample (with no ligand added) the concentration of the free metal ion [M] is equal to the total metal ion concentration $[\mathrm{M}]_{\mathrm{T}}$ and the following applies:

$$E_{\mathrm{ISE}} = E_{\mathrm{ISE}}(\mathrm{M}) = E_{\mathrm{ISE}}(\mathrm{M})_{\mathrm{T}} = E' + \frac{RT}{nF}\ln[\mathrm{M}]_{\mathrm{T}} \quad (23)$$

where E' stands for the formal potential for a specific ion sensitive probe. The actual value is of no importance in this application because we are interested in monitoring changes in the ISE response throughout a titration experiment. Hence, E' will cancel out. After the addition of a ligand, and throughout a pH titration, the response of an ISE is towards an unknown free metal ion concentration $[\mathrm{M}]_{\mathrm{pH}(i)}$ due to the formation of complexes and is given as:

$$E_{\mathrm{ISE}} = E_{\mathrm{ISE}}(\mathrm{comp})_{\mathrm{pH}(i)} = E' + \frac{RT}{nF}\ln[\mathrm{M}]_{\mathrm{pH}(i)} \quad (24)$$

Hence, we can compute the change in the ISE's response as:

$$\begin{aligned}\Delta E_{\mathrm{ISE,pH}(i)} &= E_{\mathrm{ISE}}(\mathrm{M}) - E_{\mathrm{ISE}}(\mathrm{comp})_{\mathrm{pH}(i)} \\ &= \frac{RT}{nF}\ln\left(\frac{[\mathrm{M}]_{\mathrm{T}}}{[\mathrm{M}]}\right)_{\mathrm{pH}(i)} \\ &= slope \times \log\left(\frac{[\mathrm{M}]_{\mathrm{T}}}{[\mathrm{M}]}\right)_{\mathrm{pH}(i)}\end{aligned} \tag{25}$$

where *slope* stands for the response slope of an ISE toward a change in the free metal ion concentration throughout an entire experiment. Notably, the value of *slope* must be determined experimentally as it seldom equals to the theoretical (or Nernstian) slope of 2.303 RT/nF that at 25 °C is $59.16/n$ mV.

IV. Impact of Kinetics on the Shape and Placement of Recorded Curves

Complex formation reactions require time to reach chemical equilibrium and some complexes can be formed very fast, whereas others might require a very long time. The latter case implies out-of-cell experiments, which we will not consider here. There are at least three aspects one must be aware of when studying solution equilibria by polarography, namely:

1. A single polarographic measurement requires some time and the so-called polarographic time-scale varies between techniques used. For instance, DCP operates on the time scale of seconds. This is because a single value of a step potential is applied (for 2 to 4 seconds) to a mercury drop and current is measured at the end of the life-time of the single Hg-drop. This determines the time window within which chemical (homogeneous) equilibrium must be reached at the electrode surface. DPP operates on a millisecond-scale as an additional pulse in the applied potential is imposed for 40–80 ms on the applied step potential with current being integrated at the end of the pulse. This means that chemical equilibria have a few seconds (in DCP) or a fraction of a second (in DPP) to be restored at the electrode-solution interface — this is referred to as the polarographic time scale. For homogeneous kinetics to be considered as fast, the chemical equilibria must be reached

within the time window of the polarographic technique employed. If this condition is met then such a metal–ligand system can be regarded as labile on a polarographic time-scale.

2. By reducing the metal complex(es) we "destroy" chemical equilibrium just at the working electrode surface in order to learn about equilibria in the bulk of a solution. This means that there is an interplay of several time-dependent processes taking place: time needed to re-establish chemical equilibrium *vs.* the polarographic time-scale discussed above, as well as the speed of electron transfer (heterogeneous kinetics) between the electrode surface and the species in the adjacent layer of solution. It is important to realize that the use of Eq. 17 requires fast heterogeneous kinetics, *i.e.*, the recorded polarographic signal can be seen as representing a reversible polarographic process that obeys the Nernst equation. Clearly, one must test the recorded curves for polarographic reversibility in order to eliminate curves unsuitable for the metal–ligand equilibria studies. In cases where newly formed complex(es) cannot be reduced within the available potential window, *i.e.*, prior to the reduction of water or the background electrolyte (which depends on the working electrode used, pH of the solution, kind of background electrolyte used, *etc.*) then such a metal-ligand system is called inert.
3. Complexes have to reach the electrode surface by diffusion and liberated products of the electrochemical reaction must diffuse into the bulk of the solution. This introduces a time scale needed to transport species to and from the electrode surface.

In general, electrode processes encountered in polarography of metal ions or metal complexes can be described as:

1. Reversible if the equilibrium between the oxidized and reduced forms of the species is established so rapidly that (i) the electrode potential can be described by the Nernst equation and (ii) the current is controlled by the rate of diffusion of the electroactive species to and from the electrode surface. In DCP, an important feature of a reversible electrode process is its half-wave potential $E^{\mathrm{r}}_{1/2}$. This unique potential is observed at an experimentally recorded DCP current $I = I_d/2$, where

I_d is the limiting diffusion current. $E^r_{1/2}$ occurs at a potential very near the standard potential ($E°$) for the redox couple involved, and under standard conditions $E^r_{1/2} = E°$.

2. Irreversible if slow electron transfer kinetics governs the exchange between oxidized and reduced forms of the species (heterogeneous steps whose backward rates are negligible with respect to the forward rates). The recorded signal is always shifted towards more negative potentials (as compared to that for the reversible process). However, due to a unique feature of DCP, I_d remains essentially the same regardless of the degree of reversibility.
3. Quasi-reversible processes are anything in between reversible and irreversible processes as the corresponding polarographic signals are simultaneously controlled by diffusion and an electrochemical step whose backward rate is not negligible with respect to the forward one.

One must stress that there is no rigorous mathematical description that one can apply to describe labile, non-labile, inert, *etc.* complexes. The above-mentioned classification should be used only as a descriptive guide as in reality the degree of lability or reversibility can vary a lot; hence, a very confusing picture might arise. Let us briefly discuss a few cases where the electron transfer is fast implying that the recorded curves represent electrochemically reversible process.

A. Fully Labile M–L–OH Systems

Figure 1 shows a series of reversible DP polarograms recorded on a fully labile system.[28] This is a classical picture of systems where homogeneous equilibria are established in a shorter time than the time scale of the polarographic technique used.

There are three characteristic features illustrated in this figure:

1. The DPP peak width at half the peak height is about 62 mV and remains constant for all polarograms; this is a signature of fast heterogeneous kinetics and thus electrochemical reversibility.
2. A single DPP peak shifts from a less negative to a more negative potential. This is a signature of complex formation reactions taking

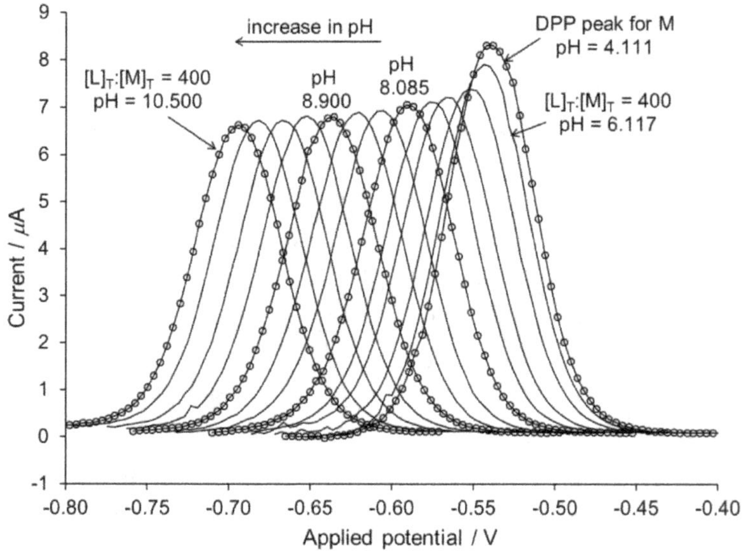

Figure 1. A set of experimental DPP curves selected from a pool of 65 polarograms recorded for the $[\text{glycine}]_T{:}[\text{Cd}]_T = 400$ at 25 °C in the background of 0.5 M (H,Na) NO_3. Initial $[L]_T = 3.3296 \times 10^{-2}$ M. Individual readings of current at the applied potential in 4 mV steps (with $E_{pulse} = 50$ mV) are shown for selected polarograms; this is to illustrate that experimental curves are not a continuous function as one might misinterpret from seeing DPP peaks as solid line traces. Unpublished data from Ref. 28.

place in a solution as the pH is increased. Hence, ML is reduced at a more negative potential than M(aq) and ML_2 requires an even more negative potential to be reduced, *etc.*

3. A relatively small decrease in the intensity of the signal is observed throughout the experiment. Typically, the drop in the intensity of a signal is due to the sample solution being diluted by addition of, *e.g.*, a NaOH solution required to adjust the pH upwards. In addition, a small decrease might be due to a decrease in diffusion coefficient D with the formation of complexes, where typically $D_{M(aq)} > D_{ML} > D_{ML_2}$ *etc.* As a rule, with an increase in the size of a complex a decrease in D is expected.

Both the intensity of a recorded signal (here the height of a polarographic peak, I_p^r) and its position along the applied potential scale (in this case the peak potential, E_p^r) are determined for each recorded polarogram; they are used by Eqs. 17 and 18. One must realize, however, that the major contribution to the corrected shift when labile systems are studied comes from the shift in the polarographic signal.

B. Inert and Non-labile Metal Complex(es)

Often, strong chelating ligands form metal complexes that are (i) non-labile — they are reduced at a much more negative potential when compared with E_p^r(M) or (ii) inert — they cannot be reduced within the available potential window. In addition, the reduction of non-labile complexes might show significant departure from electrochemical reversibility. The polydentate ligand *N,N,N′,N′*-tetra-2-picolylethylenediamine (Tpen) forms strong ML complexes with many metal ions, *e.g.*, Cd(II), Bi(III) and Pb(II).[27] These complexes are formed in very acidic medium already and either do not produce a DPP reduction peak at all or, when reduced, the recorded signals are uninterpretable in terms of providing input data for Eqs. 17 and 18. It is seen in Figure 2 that the DPP peak of

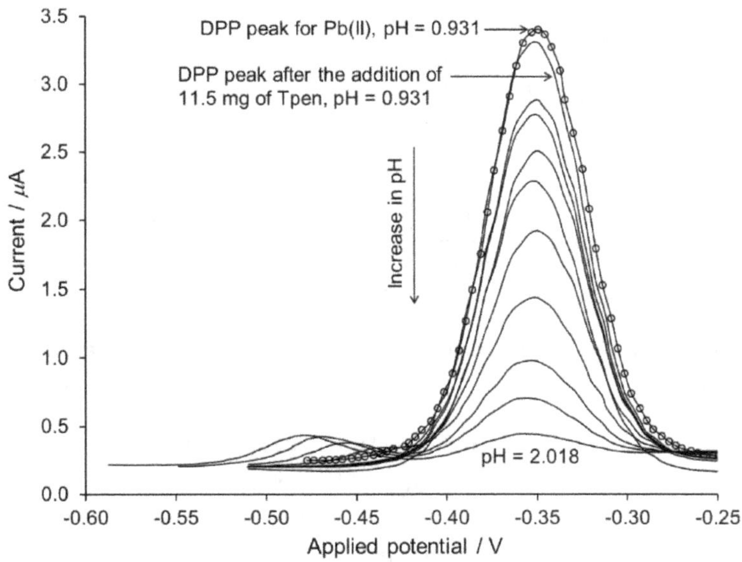

Figure 2. A set of experimental DPP curves recorded for the $[\text{Tpen}]_T$:$[\text{Pb}]_T$ = 28 at 25 °C in the background of 0.5 M (H,Na)NO_3. Initial $[L]_T = 1.3850 \times 10^{-3}$ M. Unpublished data from Ref. 27.

the free metal ion — the curve with circles — is located at about –0.35 V on the applied potentials scale. This peak, after the addition of the ligand Tpen at the pH = 0.931 – *i.e.*, 0.12 M acid solution, has slightly decreased but has not shifted along the potential scale.[27] When the pH of a solution was increased stepwise, the DPP peak at the applied potential of –0.35 V decreased further, but its peak potential E_{p} remained virtually unchanged. Knowing that formation of the $\mathrm{Pb}_p(\mathrm{OH})_r$ complexes starts only at much higher pH, it is clear that 1) the set of DPP peaks at the applied potential of about –0.35 V can only represent the uncomplexed (free) metal ion and 2) the decrease in the $I_{\mathrm{p}}(\mathrm{Pb})$ values can only be explained by the formation of lead complexes with the ligand Tpen, most likely ML. Moreover, the fact that this peak does not shift (its $E_{\mathrm{p}} = const$) on the increase in pH means that the ML complex is non-labile and it does not dissociate fast enough when free Pb(II) is reduced. In other words, we have two forms of lead cation in the solution (the free Pb(II) and PbL) and they do not exchange a metal ion on the time-scale of the DPP experiment. This is why the recorded peaks of the free metal ion have very much the same peak potential in Figure 2 as, in general, the value of $E_{\mathrm{p}}(\mathrm{M})$ does not depend on the pH of a solution, provided that hydrolysis does not take place.

This is a special case where the corrected shift defined in Eq. 18 depends only on the change in the signal intensity because $\Delta E^{\mathrm{r}}_{\mathrm{pH}(i)} = E^{\mathrm{r}}(\mathrm{M}) - E^{\mathrm{r}}(\mathrm{comp})_{\mathrm{pH}(i)} = 0$ and Eq. 17 becomes

$$\frac{RT}{nF}\ln\left(\frac{I^{\mathrm{r}}_{\mathrm{p}}(\mathrm{comp})}{I^{\mathrm{r}}_{\mathrm{p}}(\mathrm{M})}\right)_{\mathrm{x}(i)} = \frac{RT}{nF}\ln\left(\frac{[\mathrm{M}]_{\mathrm{T}}}{[\mathrm{M}]}\right)_{\mathrm{x}(i)} \tag{26}$$

C. Mixed M–L Systems Containing Labile and Non-labile/Inert Complexes

The labile part of a metal–ligand system is always used in the study of metal complexes when employing Eq. 17. This is because (i) inert complexes do not generate a polarographic signal and (ii) non-labile complexes do not shift in a predictable manner or they stop shifting; as such, they are unpredictable in terms of their position along the potential scale as well as their intensity.

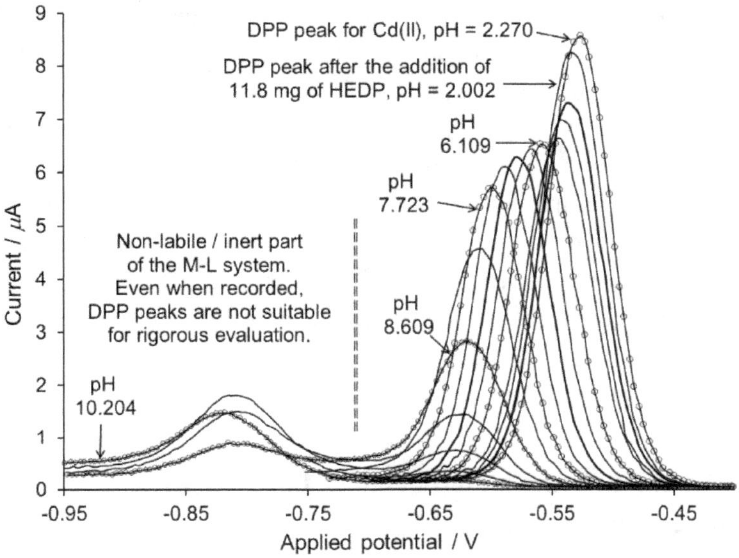

Figure 3. A set of DPP polarograms recorded for the Cd–HEDP system at a fixed $[L]_T:[M]_T = 27$. The first peak (seen at less negative potential) represents a labile part of the M–L system. Peaks at $E_{appl} < -0.75$ V represent a non-labile and essentially uninterpretable part of the metal–ligand system. Unpublished data from Ref. 26.

When the labile part of a mixed system is analyzed, then one should be able to identify two characteristic regions — see Figure 3. Initially labile metal complexes are formed (region 1, between pH 2 and 7.7). This means that a polarographic signal shifts towards potentials that are more negative and its intensity slightly decreases. In the region 2 (between pH of about 7.7 and 10), the polarographic signal attributed to the labile part of the metal–ligand system continues to shift with an increase in pH (a signature of labile complexes being present) but at the same time, the signal's intensity decreases a lot as either inert or non-labile complexes start to form and eventually dominate the composition of the solution. When labile complexes are no longer present in solution, or strictly speaking, their concentrations are too low to be detected, then the signal attributed to the labile part of the M–L system disappears entirely. In this example a minute peak representing labile complexes was still observed at pH 10.2.[26]

When labile and non-labile complexes are formed, only the polarographic signal representing all labile species present in the solution at each pH can be used in modelling of a M–L–OH system and computing formation constants. However, signals recorded in both regions contribute somewhat differently to the corrected shift term in Eq. 18. Data collected in region 1 makes a major contribution to the $CS_{pH(i)}$ from the shift in the signal, whereas increasingly an important contribution due to the relative change in the signal intensity comes from region 2. It is then very important to record a labile signal in a pH-range as wide as possible because it provides all the information needed to compute formation constants for both, labile and non-labile complexes.

D. An Example of Fast Heterogeneous but Sluggish Homogeneous Kinetics

All the above examples represent a few different trends one might expect when the appearance of polarographic signals is considered. However, the number of different scenarios emerging from polarographic studies is much larger, as electrochemical reversibility as well as homogeneous kinetics can vary widely. A set of polarograms recorded for the Cu(II)–sarcosine system[25] is shown in Figure 4, where the initial (at lower pH values) and final (at highest pH values) sets of polarograms (solid lines) appear to be reversible (or nearly so). The two sets of peaks were found to represent[25] ML and ML_2 that dominate at lower and higher pH, respectively. Notably, both ML and ML_2 complexes are reduced reversibly meaning that due to the heterogeneous kinetics being fast enough, well-shaped DPP peaks were obtained. Unfortunately, the peak at lower pH becomes wider, asymmetrical and splits into two highly overlapping peaks in the middle part in Figure 4, *i.e.*, in the pH-range of about 5–9.

This is the pH region where both complexes, ML and ML_2, are present at significant concentration levels. Polarograms recorded in the middle part (dashed lines) must be due to a homogeneous equilibrium ML $\rightleftarrows$ ML_2 being, relative to the experimental time-scale, (i) too slow to produce a single peak representing both complexes and (ii) too fast to generate two well-separated peaks representing individual ML and ML_2 complexes. Dashed line traces in Figure 4 are unsuitable for a rigorous approach in

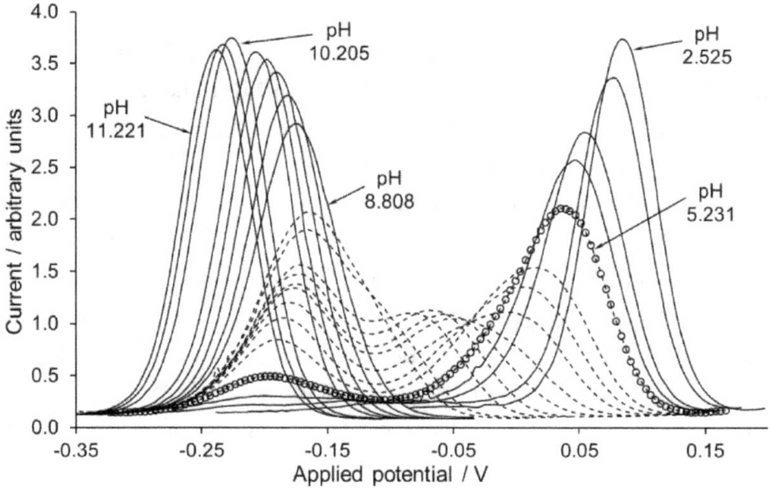

Figure 4. Selected experimental DPP curves obtained for the Cu(II)-sarcosine system at a fixed $[L]_T:[M]_T = 300$ ratio. Solid lines recorded at lowest (between 2.5 and 5.0) and the highest (between 8.8 and 11.2) pH values represent a reduction of dominant CuL and CuL_2 complexes, respectively. Adapted from Ref. 25.

modelling and refinement of formation constants. Interestingly, the initial and final solid-line polarograms provided sufficient data to generate "broken" or discontinuous complex formation curves from which reliable formation constants for the two complexes were obtained. This was achieved because E_p^r and I_p^r values representing a reversible signal were extracted from the experimental solid-line traces seen in Figure 4 using protocols described in Section XII.

To conclude, it is important to stress that employing two experimental techniques is always beneficial in arriving to a final M–L–OH model. If possible, one should consider making use of extensively tested (and experimentally verified on a large number of diverse metal–ligand systems) "classical" glass electrode potentiometry, GEP. Although, as any other analytical technique, GEP has many limitations, it still provided most of known (reported) formation constants that are regarded as reliable.[13] Using data from a well-established analytical technique is particularly important when one must verify interpretability of experimental data collected from emerging analytical techniques in terms of proposed metal–ligand models. In other words, new or modern analytical techniques that are to become "accredited" tools should be extensively tested on many M-L systems for which models and formation constants are well established and verified. Another and important point to make is that an M–L–OH model generated from a new in the field technique must not be expected to be exact copy of the one reported previously. Each analytical technique operates in a specific analytical regime ($[L]_T$, $[M]_T$, $[L]_T$:$[M]_T$ ratio, pH-range) and this usually has an impact on species formed. However, the main set of major complexes, typically ML, ML_2, *etc.*, must be reported by all analytical techniques when additionally discovered metal complexes are reported. In other words, when contradicting M–L models are proposed from two analytical techniques then results reported from at least one of them are questionable. This is because there can only be one M–L–OH model.

V. Experimental Complex Formation Curves (ECFCs)

Experienced coordination chemists might be able to predict, using accumulated knowledge and reported data for many metal-ligand systems,[11,12] major complexes that could form between a metal ion of a given oxidation state and a ligand with particular binding sites. They can also come up with reasonable guesses of the values of formation constants for the proposed complexes. This is extremely helpful but, nonetheless, it still must be confirmed (verified) by dedicated experiments. Considering less experienced researchers or newcomers to the fields of solution equilibria and coordination chemistry, they are faced with a real and difficult double challenge; namely, (i) what species can be formed, and (ii) initial guesses of their formation constants. Conveniently, however, the method of speciation described in this chapter provides simple tools to identify major and even minor complexes formed due to varied solution composition (pH or pL); this is the focus of this section.

A. Plot of Corrected Shifts ($CS_{pH(i)}$) *vs.* pH in Predicting Major $M_pL_qH_r$ or $M_pL_qOH_r$ Species Formed

Typically, many experimental curves (polarograms) are acquired on a single sample solution. In fact, one should follow a simple and general rule: the more curves recorded the more reliable the model and accurate formation constants with smaller standard deviations will be obtained. Each curve, recorded at a specific pH(i) (or pL(i)) value, is fully analyzed in order to obtain (in the case of DPP) the peak potential E_p^r(pH(i)) and peak current I_p^r(pH(i)). These experimental data are used to compute the corrected shift $CS_{pH(i)}$ according to Eq. 18. The set of $CS_{pH(i)}$ values carries necessary information about complexes formed and it is very convenient to analyze these data in the form of Experimental Complex Formation Curves (ECFCs).

An example of ECFC-pH, *i.e.*, the graph of $CS_{pH(i)}$ *vs.* pH, for the fully labile glycine–Cd(II) system is shown in Figure 5,[28] the experimental $CS_{pH(i)}$ values being shown as diamonds. In order to meaningfully interpret this highly non-linear ECFC-pH function, one must relate the observed slopes to the major form of the ligand present in a selected pH range. This is why major forms of the ligand glycine are also indicated in Figure 5 and the arrow at pH 9.54 represents the first protonation

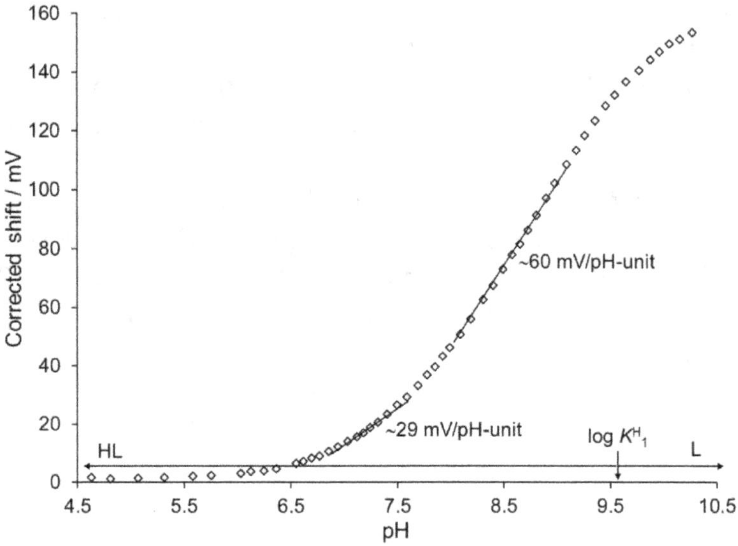

Figure 5. The experimental complex formation curve, ECFC-pH, generated by plotting along the pH scale the corrected shifts obtained for the fully labile glycine–Cd(II) system studied by DPP at $[L]_T$:$[M]_T$ ratio of 400, initial $[M]_T = 8.32 \times 10^{-5}$ M, $T = 298$ K, $I = 0.5$ M ($NaNO_3$). Adapted from Ref. 28.

constant,[11,12] as $\log K_1^{\mathrm{H}}$, of the ligand as at this pH there is an equal molar concentration of the free ligand and HL. Linking the observed slopes with a specific metal-containing species requires inputs from the theoretically predicted slopes for the assumed metal species.

To this effect, and for modelling of the $M_pL_qH_r$ complexes, the classical Nernstian *RT/nF* slope has been redefined and the new expression, called a New Nernstian Slope (*NNS*)[26] has been proposed for modelling metal–ligand systems when studied by polarography at a fixed $[L]_T$:$[M]_T$ ratio and varied pH.

$$NNS\text{-pH} = \frac{h\dfrac{RT}{nF}}{p} \quad (\mathrm{V}) \tag{27}$$

In Eq. 27, the term *NNS*-pH stands for the slope along a specific region of the $CS_{\mathrm{pH}(i)}$ *vs.* pH curve; *R*, *T*, *F*, and *n* have their common meaning, as defined in the classical Nernst equation. However, it is important to stress that *n* stands for the number of electrons involved in the reduction of a single metal ion regardless of the number of metal ions in a complex. This means that for the reduction reaction of a divalent metal ion present in a polynuclear complex, *e.g.*, $M_2L + 4e \rightarrow 2M(Hg) + L$, the value of *n* in Eq. 27 is two even though the total number of electrons needed to reduce M_2L is four.

If one prefers to express a slope in mV rather than Volts, then for an experiment performed at a room temperature of 25 °C (298.15 K) one can replace the natural logarithm ln on the right-hand side of Eq. 18 by 2.303 log (where 2.303 = log 10) and this leads to the desired expression (Eq. 28),

$$NNS\text{-pH} = \frac{h\dfrac{59.16}{n}}{p} \text{ in mV at 25 °C.} \tag{28}$$

In Eqs. 27 and 28:

1. *p* stands for the number of metal ions in $M_pL_qH_r$ or $M_pL_qOH_r$ complexes.
2. *h* represents the number of protons H required to either (i) generate the dominant H_nL form of a ligand in a solution or (ii) neutralize OH

groups that were produced (freed) from the complex due to the reduction process. Typically, h varies along the pH scale when protonated forms of L and stoichiometry of complexes change. Hence, it is expected that in many cases $h \neq r$.

For convenience, numerous examples of electrode processes involving reduction of metal complexes and accompanying post-reduction reactions taking place in a solution for any divalent metal ion containing species, *e.g.* Cd^{2+}, hence for $n = 2e$, are included in Table 1. Notably, for each entry in Table 1, the first reaction refers to a heterogeneous electron transfer at the electrode-solution interface, *i.e.*, the reduction of a specified

Table 1. Examples of a reduction process involving an assumed metal complex (top reaction) and accompanying homogeneous process (bottom reaction) taking place in the presence of a dominant H_nL form of a ligand. Parameters p and h are used to predict *NNS*-pH in mV from Eq. 28. $T = 298.15$ K and $n = 2$. Unpublished data.

No.	Reactions at a Hg electrode	H_nL	p	h	*NNS*-pH
1	$M(HL) + 2e \rightarrow M(Hg) + HL$ $HL + H \rightarrow H_2L$	H_2L	1	1	30
2	$M(HL) + 2e \rightarrow M(Hg) + HL$	HL	1	0	0
3	$M_2L + 4e \rightarrow 2M(Hg) + L$ $L + 2H \rightarrow H_2L$	H_2L	2	2	30
4	$M_2L + 4e \rightarrow 2M(Hg) + L$ $L + H \rightarrow HL$	HL	2	1	15
5	$M_2L + 2e \rightarrow 2M(Hg) + L$	L	2	0	0
6	$ML + 2e \rightarrow M(Hg) + L$ $L + 2H \rightarrow H_2L$	H_2L	1	2	60
7	$ML + 2e \rightarrow M(Hg) + L$ $L + H \rightarrow HL$	HL	1	1	30
8	$ML + 2e \rightarrow M(Hg) + L$	L	1	0	0
9	$ML_2 + 2e \rightarrow M(Hg) + 2L$ $2L + 4H \rightarrow 2H_2L$	H_2L	1	4	120
10	$ML_2 + 2e \rightarrow M(Hg) + 2L$ $2L + 2H \rightarrow 2HL$	HL	1	2	60
11	$ML_2 + 2e \rightarrow M(Hg) + 2L$	L	1	0	0

(*Continued*)

Table 1. (*Continued*)

No.	Reactions at a Hg electrode	H_nL	p	h	NNS-pH
12	$ML(OH) + 2e \rightarrow M(Hg) + L + OH$ $L + OH + 2H \rightarrow HL + H_2O$	HL	1	2	60
13	$ML(OH) + 2e \rightarrow M(Hg) + L + OH$ $OH + H \rightarrow H_2O$	L	1	1	30
14	$ML_2(OH) + 2e \rightarrow M(Hg) + 2L + OH$ $2L + OH + 3H \rightarrow 2HL + H_2O$	HL	1	3	90
15	$ML_2(OH) + 2e \rightarrow M(Hg) + 2L + OH$ $OH + H \rightarrow H_2O$	L	1	1	30
16	$M(OH) + 2e \rightarrow M(Hg) + OH$ $OH + H \rightarrow H_2O$	—	1	1	30
17	$M(OH)_2 + 2e \rightarrow M(Hg) + 2OH$ $2OH + 2H \rightarrow 2H_2O$	—	1	2	60

metal-containing complex. Here the "freed" atom of a metal leaves the aqueous solution and dissolves in a mercury drop — an amalgam M(Hg), or a solution of M in liquid mercury, is formed. The second reaction describes a post-electrochemical reduction homogeneous reaction in aqueous solution. This reaction involves protons that are used to produce an assumed dominant form of a ligand H_nL at a specific pH value of a solution and/or neutralize the freed OH groups.

Reactions involving protons are very fast and, in principle, can be regarded as nearly instantaneous (kinetically facile) processes producing an H_nL form of a ligand and/or H_2O.

To aid understanding of how to use the data in Table 1 in modelling operations, let us interpret the slopes found for the ECFC-pH in Figure 5:

1. A slope of zero mV/pH in the pH-range between 4.5 and 6.0 is observed where HL is the dominant form of glycine. This slope suggests two possibilities, either:
   (a) Cd(II) and glycine do not form a complex in this pH-range at all. Hence, the recorded polarogram must represent the reduction of a free metal ion (a Nernstian reduction of a free metal ion occurs at a specific and fixed potential, regardless of pH), or

(b) A Cd(HL) complex is formed in solution ($p = 1$) and when reduced the HL form of the ligand is released — see entry 2 in Table 1. As HL is the predominant form in this pH-range, the protonated HL ligand recovered from the reduction of the Cd(HL) complex will not be involved in a reaction with a proton H ($h = 0$). Substituting $p = 1$, $h = 0$ and $n = 2$ into Eq. 28 one obtains *NNS*-pH = 0 mV/pH-unit; clearly, also in this case no shift in the recorded signal would be observed.

2. A slope of 29 mV/pH-unit is marked at pH of about 7.5 ± 0.5, where HL dominates all ligand species. Hypothetically, this slope can be attributed to a presence of:
   (a) the M(OH) complex ($p = 1$; still no complex with glycine is formed) — see entry 16 in Table 1 that shows recombination of freed OH ions with one proton H ($h = 1$) to form a water molecule, or
   (b) the ML complex ($p = 1$) — see entry 7 in Table 1 that also shows the need of one proton ($h = 1$) to form HL.
   Hence, both processes require a single proton ($h = 1$) per individual chemical event ($H + OH \rightleftarrows H_2O$ and/or $H + L \rightleftarrows HL$ and for both $p = 1$ and $n = 2$); substituting these parameters into Eq. 28 gives *NNS*-pH ≈ 29 mV/pH-unit. Note that in this pH-range the experimental points vary exponentially rather than linearly. This indicates that neither M(OH) nor ML is the major metal-containing complex in solution.
3. A slope of 60 mV/pH-unit between pH of 8 and 9 is observed where HL still dominates the ligand species. This might indicate formation of a consecutive complex due to:
   (a) further hydrolysis of the metal ion; hence, one should consider the $M(OH)_2$ formation ($p = 1$ as shown for entry 17 in Table 1) that upon reduction would release two OH ions that require 2 protons ($h = 2$) to form two H_2O molecules; or
   (b) coordination of a second molecule of the ligand glycine, giving the ML_2 complex (see entry 10 in Table 1: $p = 1$) where liberated ligand molecules require 2 protons in order to change to the major HL form of the ligand.

   In both cases, we have $p = 1$, $h = 2$ and $n = 2$; this is the case where Eq. 28 predicts the observed slope of about 60 mV/pH-unit.

4. In the pH region of the free (fully deprotonated) ligand (above pH of about 9.5), a gradual drop in the slope is observed. This is an important observation as it allows one to eliminate the formation of M(OH) and $M(OH)_2$ complexes as major species in a solution. This is because the presence of the $M(OH)_2$ complex as a major species would always produce a constant slope of 60 mV/log-unit in the pH-range irrespective of the dominant form of the ligand. Moreover, it is also known that Cd(II) can form $M(OH)_3$ and $M(OH)_4$ complexes. If metal hydrolysis were to dominate in solution, then formation of $M(OH)_2$ would have to be followed by formation of $M(OH)_3$ and $M(OH)_4$ (hydrolyzed complexes that would produce slopes of 90 and 120 mV/pH-unit, respectively). Note that this scenario is a possibility only when no precipitation of $Cd(OH)_2(s)$ occurs. From the above discussion, it is obvious that the observed trend in the corrected shift function at the highest pH values eliminates hydrolysis as a major process in the solution. Moreover, precipitation of $Cd(OH)_2(s)$ did not occur as it would remove Cd^{2+} ions from the bulk solution. Consequently, no reduction peak should have been recorded resulting in no experimental points, but data shown in Figure 5 contradicts such an assumption entirely.

Clearly, all the above shows that Cd–glycine complexes are formed and one would start building an initial M–L–OH model containing two major and confirmed complexes from the above modelling — CdL and CdL_2. Naturally, one would bear in mind a possibility of the Cd(HL) formation and necessity of inclusion of all known $Cd_p(OH)_r$ species in the initial model.

B. Plot of Corrected Shifts ($CS_{pH(i)}$) *vs.* log [L] in Predicting Major M_pL_q Species Formed

To make the relationship between the corrected shift $CS_{pH(i)}$ *vs.* log [L] applicable in studying M_pL_q complexes (hence also to polynuclear metal complexes) at any pH, the classical Nernstian *RT*/*nF* slope must be modified to:

$$NNS\text{-pL} = \frac{q\dfrac{RT}{nF}}{p} \text{ (in V) and } NNS\text{-pL} = \frac{q\dfrac{59.16}{n}}{p} \text{ (in mV at 25 °C)} \quad (29)$$

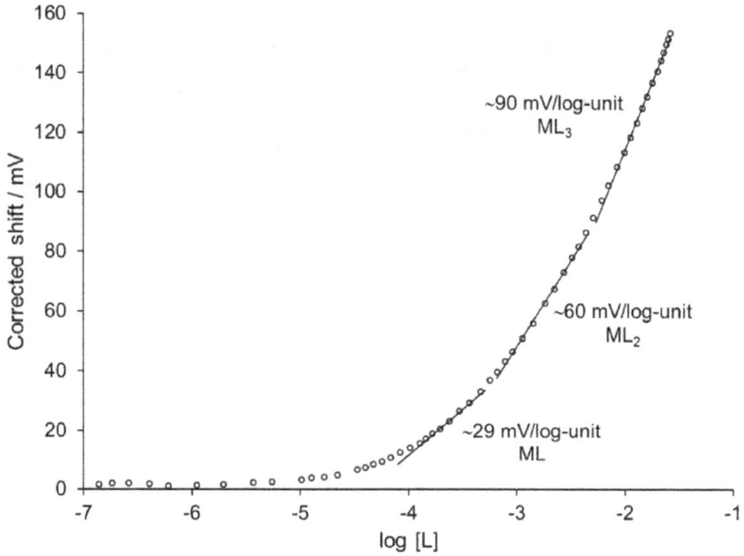

Figure 6. The experimental complex formation curve, ECFC-pL, generated by plotting corrected shift, *i.e.*, $CS_{pH(i)}$ values, *vs.* computed the free ligand concentration. Data were obtained for the fully labile glycine–Cd(II) system studied by DPP at $[L]_T$:$[M]_T$ ratio of 400, initial $[M]_T = 8.32 \times 10^{-5}$ M, $T = 298$ K, $I = 0.5$ M ($NaNO_3$). Experimental data from Ref. 28 were used to prepare this figure.

The *NNS*-pL term stands for the new Nernstian slope applicable only to the M_pL_q complexes when analyzed using the corrected shift $CS_{pH(i)}$ as shown in Figure 6. Notably, the same experimental data from Ref. 28 where used to generate Figures 5 and 6 where the $[L]_T$:$[M]_T$ ratio of 400 was used. Working with a large excess of a ligand allows one to compute reasonable the free ligand concentration at each experimental point by solving the MBE for the total ligand concentration (Eq. 22) without including any metal-containing species, in this case $[L]_T = [L] + [HL] + [H_2L]$. Importantly, such simplified approach has no impact on interpretation of slopes seen in Figure 6. To aid interpretation of slopes in Figure 6 one can make use of data included in Table 2.

The analysis of trends shown in Figure 6 leads to the following conclusions:

1. The overall shape of the ECFC-pL = $CS_{pH(i)}$ *vs.* log [L] function can be seen as exponential in nature. This implies that the M_pL_q complexes

Table 2. Examples of a reduction process involving an assumed metal complex (top reaction) and accompanying homogeneous process (bottom reaction) taking place in the presence of a dominant H_nL form of a ligand. Parameters p and q are used to predict *NNS*-pL in mV from Eq. 29. $T = 298.15$ K and $n = 2$. Unpublished data.

No.	Reactions at a Hg electrode	H_nL	p	q	*NNS*-pL
1	$M_2L + 4e \rightarrow 2M(Hg) + L$ $L + 2H \rightarrow H_2L$	H_2L	2	1	15
2	$M_2L + 4e \rightarrow 2M(Hg) + L$ $L + H \rightarrow HL$	HL	2	1	15
3	$M_2L + 2e \rightarrow 2M(Hg) + L$	L	2	1	15
4	$ML + 2e \rightarrow M(Hg) + L$ $L + 2H \rightarrow H_2L$	H_2L	1	1	30
5	$ML + 2e \rightarrow M(Hg) + L$ $L + H \rightarrow HL$	HL	1	1	30
6	$ML + 2e \rightarrow M(Hg) + L$	L	1	1	30
7	$ML_2 + 2e \rightarrow M(Hg) + 2L$ $2L + 4H \rightarrow 2H_2L$	H_2L	1	2	60
8	$ML_2 + 2e \rightarrow M(Hg) + 2L$ $2L + 2H \rightarrow 2HL$	HL	1	2	60
9	$ML_2 + 2e \rightarrow M(Hg) + 2L$	L	1	2	60

are formed in a consecutive manner and none are dominant species. Consequently, a mixture of M_pL_q complexes must exist at any pH of a solution.

2. It is important to note that the free ligand concentration varied in quite a large range, between 1.22×10^{-7} and 2.64×10^{-2} M (log [L] of –6.914 and –1.579, respectively), whereas the initial total ligand concentration was 3.33×10^{-2} M and did not change much throughout the experiment.
3. A slope of about 29 mV/log-unit is observed in Figure 6 at log [L] of about –3.5. From Table 2, a slope of about 30 is attributed to the ML formation regardless of the dominant form of ligand — see entries 4–6. Note that (i) the ML complex was also identified from the analysis of ECFC-pH in Figure 5, where a slope of about 30 mV/pH-unit is observed at pH about 7.5, where HL is a dominant form of glycine

and (ii) the same experimental points support formation of ML on the ECFC-pH (Figure 5) and ECFC-pL (Figure 6) graphs.

4. When $n = 2$, $p = 1$ and $q = 2$ a slope of about 59 mV at 25 °C is predicted for the ML_2 complex by Eq. 29. Formation of ML_2 is fully confirmed by the slope of about 60 mV/log-unit seen in Figure 6 at log [L] values between –3 and –2.2 that is equivalent to the pH-range between 8 and 8.9, where a similar slope is seen in Figure 5.
5. A slope of about 90 mV/pH-unit is observed in Figure 6 for the largest free ligand concentrations, *i.e.*, at the highest pH values. Clearly, such a slope is predicted by Eq. 29 for ML_3 ($n = 2$, $p = 1$ and $q = 3$) and this perfectly completes a sequence of the consecutive ML, ML_2 and ML_3 complexes formation. Note that the ML_3 formation could not be predicted from trends in Figure 5 because it becomes a major metal containing species only in the pH-range where the free ligand L dominates. This nicely explains why a gradual drop in $CS_{pH(i)}$ *vs.* pH is observed in Figure 5, starting from pH = 9.54, as no H-ion is required to protonate L released from the reduction of ML_3. Clearly, the observed slope is an interplay between two major processes that follow the reduction reaction $ML_3 + 2e \rightarrow M(Hg) + 3L$. The first process generates 0 mV/pH-unit slope when the ligand liberated from the complex remains as L. The second process L + H → HL on its own would generate a slope of 90 mV/pH-unit as three free ligand molecules formed from the ML_3 reduction would require 3H to form dominant HL form. Hence, the slope decreases when a fraction of the protonated ligand HL decreases with an increase in pH.

Clearly, the analysis of trends observed for the ECFC-pL proved to be very useful, as it not only confirmed the formation of ML and ML_2 predicted from the trends identified for the ECFC-pH, but also revealed the presence of ML_3.

VI. Protocol to Compute Excellent Estimates for Formation Constants

We have indisputably established the formation of three Cd(II) complexes with glycine. In the next step we have to predict their formation constants and our estimates must be quite good, otherwise the refinement operations might fail. In this section, we will briefly describe how one can compute "excellent" initial values of formation constants that are suitable as inputs for refinement operations.

It has been shown[17] that when a complex formation reaction:

$$M + qL + r(OH) \rightleftarrows ML_q(OH)_r \tag{30}$$

is considered and an experiment is performed as an acid-base titration, then the following equation holds at any pH of the solution:

$$\begin{aligned} E^{\mathrm{r}}_{\mathrm{p}}(\mathrm{M}) - E^{\mathrm{r}}_{\mathrm{p}}(\mathrm{comp})_{\mathrm{pH}(i)} - \frac{RT}{nF}\ln\left(\frac{I^{\mathrm{r}}_{\mathrm{p}}(\mathrm{comp})}{I^{\mathrm{r}}_{\mathrm{p}}(\mathrm{M})}\right)_{\mathrm{pH}(i)} &= \\ \frac{RT}{nF}\ln\beta_{\mathrm{ML}_q(\mathrm{OH})_r} + q\frac{RT}{nF}\ln[\mathrm{L}]_{\mathrm{pH}(i)} + r\frac{RT}{nF}\ln[\mathrm{OH}]_{\mathrm{pH}(i)} & \end{aligned} \tag{31}$$

This can also be written in terms of the corrected shift as:

$$CS_{\mathrm{pH}(i)} = \frac{RT}{nF}\ln\beta_{\mathrm{ML}_q(\mathrm{OH})_r} + q\frac{RT}{nF}\ln[\mathrm{L}]_{\mathrm{pH}(i)} + r\frac{RT}{nF}\ln[\mathrm{OH}]_{\mathrm{pH}(i)} \tag{32}$$

Eq. 32 can be conveniently used to estimate the formation constant of a selected $ML_q(OH)_r$ complex. An excellent estimate will be obtained provided that the complex $ML_q(OH)_r$ is the predominant species present in a solution at selected pH(i). When protonated complexes ML_qH_r are formed (at lower pH values), to compute the estimates in their formation constants, the [OH] term in Eqs. 31and 32 must be replaced by [H]. Focusing on ML_q complexes implies that $r = 0$ and Eq. 32 becomes:

$$CS_{\mathrm{pH}(i)} = const + q\frac{RT}{nF}\ln[\mathrm{L}]_{\mathrm{pH}(i)} \tag{33}$$

where $const = \frac{RT}{nF}\ln\beta_{\mathrm{ML}_q(\mathrm{OH})_r}$. Note that the constant value (*const*) depends on the temperature at which the experiment is performed as well as on the ionic strength of the solution. From Eq. 33 it follows that, regardless of the pH-range considered, *i.e.*, regardless of the H_nL form of a ligand dominating the composition of a solution, there should be a linear relationship between the corrected shift $CS_{\mathrm{pH}(i)}$ in the recorded polarographic signal and the molar concentration of the free ligand [L].

To achieve our goal we will make use of trends in Figures 5 and 6 in Section V and Eq. 32 that can be re-arranged to a suitable form for the purpose. As we are dealing with ML_q complexes in this example, the last term in Eq. 32 is zero as $r = 0$ and for this specific case (temperature = 25 °C and $n = 2$) we can write:

$$CS_{\mathrm{pH}(i)} = 29.58 \log \beta_{\mathrm{ML}_q} + q29.58 \log [\mathrm{L}]_{\mathrm{pH}(i)} \text{ (in mV)} \quad (34)$$

Furthermore, it is more convenient to compute $\log \beta_{\mathrm{ML}_q}$ rather than the β_{ML_q} value. We can easily rearrange Eq. 34 to give a form suitable for a single point calculation, namely:

$$\log \beta_{\mathrm{ML}_q} = \{CS_{\mathrm{pH}(i)} - q29.58 \log [\mathrm{L}]_{\mathrm{pH}(i)}\} / 29.58 \quad (35)$$

Clearly, we need only two values, $CS_{\mathrm{pH}(i)}$ and $\log [\mathrm{L}]_{\mathrm{pH}(i)}$, in order to obtain $\log \beta_{\mathrm{ML}_q}$. The values of corrected shifts come directly from an experiment. Regarding the free ligand concentration, we can ignore all terms in Eq. 22 that contain concentrations of metal species because we typically work with a significant excess of ligand. Hence, a simplified form (Eq. 36) can be used as (i) it is valid at any pH of a solution, (ii) $[\mathrm{L}]_\mathrm{T}$ and [H] are known from an experiment, and (iii) the protonation constants for the ligand are also known or must be determined prior a polarographic experiment.

$$[\mathrm{L}]_\mathrm{T} = [\mathrm{L}] + \beta_1^\mathrm{H}[\mathrm{H}][\mathrm{L}] + \beta_2^\mathrm{H}[\mathrm{H}]^2[\mathrm{L}] = [\mathrm{L}](1 + \beta_1^\mathrm{H}[\mathrm{H}] + \beta_2^\mathrm{H}[\mathrm{H}]^2) \quad (36)$$

One must realize that Eq. 35 applies to a single metal complex. From this it follows, that one must select experimental data at pH-values at which the largest concentration of a selected complex is expected. To achieve that one can select points in the middle of the trends (slopes) marked in Figures 5 and 6 in Section V. Data selected for computing estimates of formation constants is included in Table 3. Although there is no need to use two or more sets of data, constancy in the obtained log values for ML, ML_2 and ML_3 provides assurance that data points at the most

Table 3. Corrected shifts and concentrations of a free ligand computed at the selected pH values. They were used in Eq. 35 to calculate the overall concentration formation constants, as log β values, for ML, ML_2, and ML_3 complexes. Table was prepared using data from the study of the Cd(II)–glycine system by DPP using an acid-base titration protocol.[28] Initial concentrations of $[L]_T$ and $[M]_T$ were 3.330×10^{-2} M and 8.320×10^{-5} M, respectively.

pH	*CS*/mV	log [L]	log β
ML			
7.317	23.245	–3.705	4.49
7.403	25.802	–3.620	4.49
ML_2			
8.581	80.582	–2.488	7.70
8.651	84.921	–2.425	7.72
ML_3			
9.651	138.369	–1.745	9.91
9.769	142.207	–1.699	9.90

suitable pH values were selected. In other words, the %-fractions of the relevant complexes formed are close to the largest values at the selected pHs, otherwise calculated log values would differ significantly more.

Finally, it is appropriate to comment on the expected quality of the log β_{ML_q} estimates. In general, the estimates will be closer to the refined values when a selected complex is indeed a major species in a solution. This is because the experimentally obtained $CS_{pH(i)}$ term in Eq. 35 is attributed just to a single complex for which the formation constant is calculated. In reality, all metal containing species contribute to the shift and intensity of recorded polarograms. Hence, just a fraction of the experimentally obtained $CS_{pH(i)}$ value can be linked with a specific complex. It is then very clear that Eq. 35 will always produce somewhat larger values than those refined from solving mass-balance equations.

The approach described is applicable to any system studied by polarography. In many instances, however, there would not be well defined slopes on the ECFC-pH or ECFC-pL graphs and this is particularly true when many complexes are present in a solution at a specific pH value. In such a case one must identify experimental points for which trends in $CS_{\mathrm{pH}(i)}$ *vs.* either pH or pL (or log [L]) asymptotically approach meaningful values and data in Tables 1 and 2 (Section V) can be used to guide the selection. In other words, for $n = 2$ one would select data at pH values for which slopes approach 15, 30, 60, 90, *etc.*, mV/per a pH (log) unit.

VII. Computed Complex Formation Curve (CCFC) and Its Use in Modelling of a M–L–OH System and Refinement of Formation Constants

The computed complex formation curve, CCFC, is obtained by plotting values computed from the right-hand side of Eq. 17 (corrected shifts *vs.* pH). One can generate the CCFC for different purposes, namely:

1. As a very useful tool in designing optimal experimental conditions for any assumed M–L–OH model and formation constants prior to running any experiment. To this effect, one can test the impact of the total metal ion concentration and the $[\mathrm{L}]_\mathrm{T}$:$[\mathrm{M}]_\mathrm{T}$ ratio on the expected corrected shift. Naturally, the corrected shifts $CS_{\mathrm{pH}(i)}$ values are to be measured and it is always a good analytical practice to strive to generate an experimental signal as large as possible. Furthermore, one can also vary the assumed formation constants and assess their impact on the values of corrected shifts as well as predict at what pH a signal should start shifting due to the formation of complexes. This will be illustrated in Section IX.
2. To assess and understand the interplay between the experimentally recorded corrected shifts, say from the first exploratory acid–base titration, and the formation constants of assumed metal complexes. One can also analyze the impact (if any) of hydrolysis constants as well as precipitation (controlled by a solubility product K_{sp} of hydroxo metal species) on corrected shifts.
3. To refine formation constants of major metal complexes identified from the analysis of slopes along the experimental complex formation curve. After refinement operations, one can identify pH-ranges where the CCFC does not fit the experimental points well, *i.e.*, the ECFC. If

such regions exist then it is possible to come up with reasonable guesses of possible complexes that might improve the overall fit of the CCFC to the ECFC. One must realize that not all complexes formed in solution can be identified directly from trends in the ECFC. This is because, more often than not, the experimentally observed slopes represent a mixture of major and minor species present simultaneously in a solution. In other words, the theoretical Nernstian slopes will only be found on the ECFC when one metal species dominates in a wide pH-range (at least 2–3 pH-units).

Regardless of the above three scenarios, to generate a CCFC one needs a M–L–OH model that is as complete as possible; a set of metal containing species as well as ligand containing species together with formation/protonation constants for all of them are required. Obviously, some formation constants that must be obtained from dedicated experiments (*e.g.*, all known $M_p(OH)_r$ as well as protonation constants for a ligand) are kept fixed (unchanged) during computing CCFCs and refinement operations. In order to demonstrate the usefulness and power of CCFCs, let us briefly illustrate and discuss above-metioned scenarios 2 and 3. To this effect, we will use experimental data obtained for the Cd–glycine–OH system from a DPP experiment at the $[L]_T$:$[M]_T$ ratio of 400, with initial $[M]_T = 8.32 \times 10^{-5}$ M, and ionic strength 0.5 M $(H,Na)NO_3$. Protonation constants for glycine, the dissociation constant for H_2O, the overall formation constants for the $Cd_p(OH)_r$ hydroxide complexes included in modelling are reported elsewhere.[28]

A. Examples of CCFCs Used in Preliminary Modelling

Relevant computed corrected shifts that were selected from a preliminary experiment performed in a wide pH range are plotted as circles in Figure 7a. The generated ECFC-pH reveals that the polarographic signal started to shift significantly only from pH 6 onwards. The S-shape of the ECFC-pH trend is a signature of consecutive formation of complexes that form in the environment dominated by a singly protonated HL form of the ligand glycine. The arrow perpendicular to the X-axis marks the first

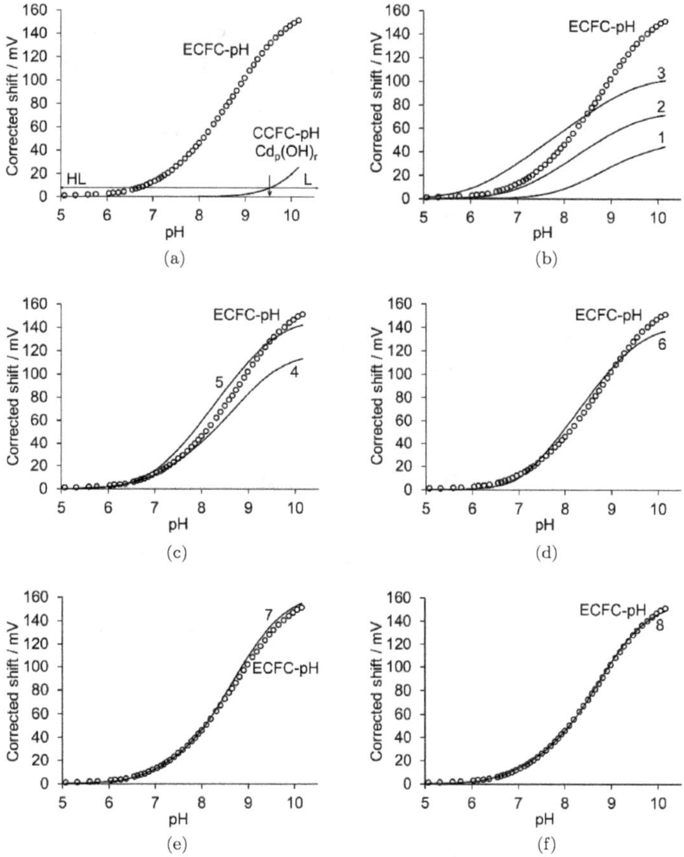

Figure 7. Modelling of a solution composition by use of CCFCs (solid lines) using the ECFC-pH (circles) obtained for the Cd(II)–glycine aqueous solution at 25 °C, ionic strength I = 0.5 M (H,Na)NO_3, the $[L]_T$:$[M]_T$ concentration ratio of 400, the total metal ion concentration $[M]_T = 8.32 \times 10^{-5}$ M. The CCFCs were obtained for: all known $Cd_p(OH)_r$ complexes (part a); $Cd_p(OH)_r$ plus ML with log β_1 set to 3, 4 and 5 — curves 1, 2 and 3, respectively (part b); $Cd_p(OH)_r$, ML with log β_1 set to 4.2 plus ML_2 with log β_2 set to 7 and 8 — curves 4 and 5, respectively (part c); the M–L–OH system represented by curve 5 was refined and produced curve 6 obtained for all $Cd_p(OH)_r$ species plus ML with refined log β_1 = 3.83 and ML_2 with refined log β_2 = 7.83 (part d); $Cd_p(OH)_r$ plus log β_1 set to 4.2, log β_2 set to 7.2 and ML_3 set to 10 — curve 7 (part e); Model represented by line 7 was refined and the CCFC seen as line 8 (part f) represents all $Cd_p(OH)_r$ species plus ML with refined log β_1 = 4.26, ML_2 with refined log β_2 = 7.17 and ML_3 with refined log β_3 = 9.80. Unpublished data except experimental points coming from Ref. 28.

protonation constant, $\log \beta_1^{\mathrm{H}} = 9.58$; notably, there is an equal molar concentration of HL and L forms of the ligand at pH = 9.58. To assure that the observed shift is not caused just by cadmium complexes with OH^-, a theoretical CFC was computed only for the $Cd_p(OH)_r$ species — see the solid line in Figure 7a. As such, the CCFC curve obtained for the $Cd_p(OH)_r$ species must be seen as hypothetical. This is because, in the absence of Cd_pL_q complexes, precipitation of $Cd(OH)_2(s)$ has been predicted already at pH of 8.7 for $\log K_{sp} = -14.3$ and the experimental value of $[M]_T = 8.3 \times 10^{-5}$ M. This means that no polarographic signal would be observed as the $Cd^{2+}(aq)$ ions would essentially be removed from solution due to precipitation.

To better understand how the formation of the ML_q complexes contributes to the corrected shift, a M–L–OH system was simulated that initially only incorporated CdL and all $Cd_p(OH)_r$ species. Solid line traces marked 1, 2 and 3 in Figure 7b represent such systems with $\log K_{ML}$ ($\log \beta_1$) set to 3, 4 and 5, respectively. Analysis of the observed trends in the CCFCs leads to a few important observations, namely:

1. It is clear that $\log K_{ML}$ must be slightly larger than 4 as the CCFC shown in Figure 7b as line 2 is the only one that asymptotically approaches the ECFC at lower pH values. Note that CCFCs marked as lines 1 and 3 are significantly below and above the experimental points at lower pH values, respectively.
2. The M–L–OH system must also contain other ML_q complexes as line 3 (computed for $\log K_{ML}$ set to 5) cannot reproduce the shape of the ECFC-pH; line 3 predicts too large $CS_{\mathrm{pH}(i)}$ values at low pHs and still is not able to reach the experimental trend at higher pH values.

Using the conclusions arrived at from the analysis of trends in Figure 7b, $\log K_{ML}$ was set to 4.2 and the ML_2 complex was added to the M–L–OH model, as it is the most likely complex one expects to form after ML. Two CCFCs, seen in Figure 7c as solid lines 4 and 5, were obtained by setting the values of $\log \beta_2$ to 7 and 8, respectively. It is evident that $\log \beta_2$ must be somewhat larger than 7 but still smaller than 8 as curve 5 is above the ECFC-pH in a wide pH-range. Just as a test, the formation constants for ML and ML_2 were refined and the resultant CCFC (seen as solid line 6 in Figure 7d) definitely confirmed that another complex, most likely ML_3, is required to better reproduce the ECFC-pH.

The solid line 7 seen in Figure 7e represents a CCFC obtained for all $Cd_p(OH)_r$ species plus ML (log β_1 set to 4.2), ML_2 (log β_2 set to 7.2) and ML_3 (log β_3 set to 10.0). This line follows the overall trend of experimental data and is very close to the circles. These formation constants proved to be perfect inputs for refinement operations that produced solid line 8 in Figure 7f with refined log $\beta_1 = 4.26 \pm 0.02$, log $\beta_2 = 7.17 \pm 0.04$, and log $\beta_3 = 9.80 \pm 0.05$. The CCFC (solid line 8) in Figure 7f fits most of experimental points very well. However, a careful inspection reveals two pH-regions, *i.e.*, for the lowest and highest pH values, where solid line 8 is below the ECFC-pH. This could indicate that some additional Cd(II) complexes with glycine might be present in solution and this will be discussed in the following section.

B. CCFCs Used in Final Modelling

Figure 7 shows just a small sample of how CCFCs might be employed in the study of solution equilibria. It also provides an insight on how one can (i) explore the impact of experimental conditions (*e.g.*, the total metal ion and ligand concentrations) or (ii) select a specific pH-range for collecting experimental points, *etc.*, by computing purposely selected CCFCs that are needed for fine-tuning experimental conditions. Moreover, one can also arrive at the final M–L–OH model using protocols illustrated in Figure 7 and, although such an approach is very informative, it takes a lot of time.

If arriving at the final model and best formation constants are of interest, then one should make use of estimates in formation constants obtained from protocols described in Section VI. It must be stressed again that in computing complex formation curves all known $M_p(OH)_r$ species must be included. The experimental complex formation curve (ECFC-pH, circles — experimental data from Ref. 28), the complex formation curve computed for the initial estimates of formation constants for ML, ML_2 and ML_3 (CCFC-pH-1, solid line) and the CCFC-pH-2 obtained after refinement of the estimates for the same M–L–OH model (dashed line) are seen in

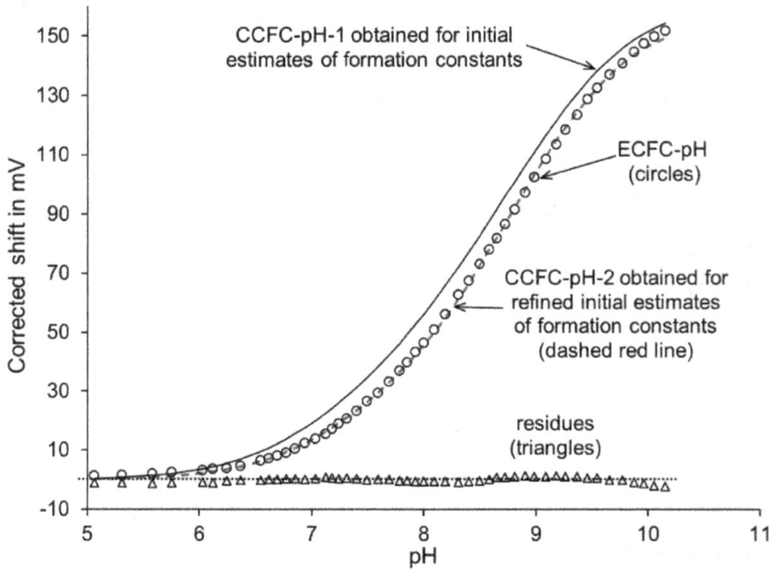

Figure 8. The ECFC-pH (circles) obtained for the Cd(II)–glycine aqueous solution — see Figure 7 caption for experimental details. Solid line is a CCFC-pH-1 obtained for estimates log $\beta_1 = 4.49$, log $\beta_2 = 7.70$, and log $\beta_3 = 9.90$ (they are included in Table 3). Dashed red line (CCFC-pH-2) was obtained after refinement of estimates for the ML, ML_2 and ML_3 complexes that produced log $\beta_1 = 4.26$, log $\beta_2 = 7.17$ and log $\beta_3 = 9.80$, respectively. Triangles represent residues that show the systematic deviations of the CCFC-pH-2 from ECFC-pH data (circles). Unpublished data. Experimental points are from Ref. 28.

Figure 8. There are at least three important points to make when looking at the curves in Figure 8:

1. The solid line, representing the CCFC-pH-1 obtained for the estimated formation constants, follows the trend in experimental points (circles) very well. At the same time, the solid line is consistently above the experimental points except at the lowest pH values. This is a common feature of the CCFC when estimates in formation constants obtained from Eq. 35 are used — too large values in the computed corrected shifts observed are the product of estimates in formation constants being too large; recall that each slope seen in Figure 6 (the experimentally recorded corrected shift) was attributed to just a single complex.

2. The dashed line, representing the CCFC-pH-2 obtained after the refinement operations, follows most of experimental points very well. The refined formation constants for ML, ML_2, and ML_3, namely log $\beta_1 = 4.26$, log $\beta_2 = 7.17$, and log $\beta_3 = 9.80$ are a fraction of log unit smaller when compared with initial estimates (Table 3). This not only demonstrates how useful simple protocols described in Sections V and VI are, but also the results shown in Figure 8 support protocols employed to identify major metal complexes.
3. CCFC-pH-2 is below the experimental points at the lowest and the highest pH values. This means that some complexes are missing in the initial model and this is clearly seen from residues (triangles), *i.e.*, differences between individual points of the CCFC-pH-2 and ECFC-pH data.

In order to improve fitting of the CCFC to the ECFC, the M(HL) and $ML_2(OH)$ complexes were included in the model. The former complex was selected to improve the fit at the lowest pH values, as this is the most likely metal-containing species that can start forming prior ML. To improve the fit at the highest pH values, $ML_2(OH)$ was selected. This is because (i) ML_4 is unlikely to form due to the most common coordination number for Cd(II) being six, and (ii) one might expect the formation of hydrolyzed complexes at higher pH values. Any ligand must compete with OH^- for a metal ion; the higher the pH the more favorable the environment for coordinating hydroxide ions. In fact, one should also consider the solubility product K_{sp} for $Cd(OH)_2(s)$ to make sure that precipitation does not occur.

As one can see in Figure 9,[28] the CCFC-pH (shown as a dotted line) obtained for the final M–L–OH model fits the experimental data extremely well. The final model is made of five Cd(II) complexes with glycine, M(HL), ML, ML_2, ML_3, and $ML_2(OH)$, for which log β values were refined as 10.07 ± 0.10, 4.20 ± 0.02, 7.30 ± 0.03, 9.64 ± 0.04, and 11.49 ± 0.08, respectively.

Eq. 25 in Section III describes the <u>change</u> in the ISE's response (potential) towards changes in a solution composition when new metal complexes are formed. Notably, Eq. 25 treats a potentiometric probe

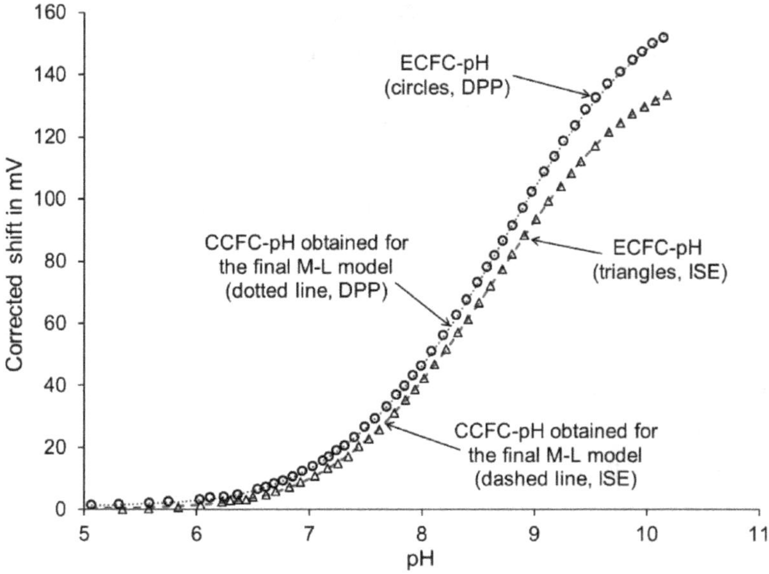

Figure 9. Experimental (circles) and calculated (dotted line) CFCs obtained from DPP as well as experimental (triangles) and calculated (dashed line) CFCs obtained from Cd-ISE at $[L]_T:[M]_T$ ratios of 400 (DPP) and 100 (ISE). $[M]_T$ was 8.320×10^{-5} M (DPP) and 2.495×10^{-4} M (ISE). Adapted from Ref. 28.

(sensor) in the same way as polarographic data are treated when no change in signal intensity (I_p in DPP or I_d in DCP) occurs. In such a case, the corrected shift becomes equal to the change in ISE potential and when plotted *vs.* pH it should produce the experimental complex formation curve. A full confirmation of this conclusion is seen in Figure 9 where triangles represent the ISE-generated ECFC-pH obtained for the Cd(II) – glycine–OH system. There is a striking similarity between the ECFCs obtained from such different experimental techniques; dynamic and non-equilibrium polarography (circles) and equilibrium potentiometry (triangles).

The only visible difference is in somewhat smaller values in the corrected shift observed from ISE-based measurements. This is because the $[L]_T:[M]_T$ ratios were 400 (for DPP) and 100 (for ISE). One must realize that the larger the $[L]_T:[M]_T$ ratio, the higher the mole fraction of complexes formed at a specific pH. Thus, the polarographic ECFC is above that obtained from potentiometry. Most importantly, the CCFC that fits ISE-generated experimental points best (triangles in Figure 9) was obtained for exactly the same M–L–OH model as from polarograhy,[28] *i.e.*, M(HL), ML, ML_2, ML_3, and $ML_2(OH)$ for which log β values were refined from the ISE experiment as 9.93 ± 0.18, 4.18 ± 0.03, 7.44 ± 0.04, 9.67 ± 0.06, and 11.29 ± 0.13, respectively.

It is then clear that experimental and computed CFCs can be generated from dynamic and equilibrium techniques. This is very promising, as the same modelling operations can be employed in search of the best experimental conditions suitable for both techniques. Moreover, the same software, originally designed and developed for the treatment of voltammetric data, can be used in the refinement of formation constants using data from a potentiometric sensor that, when Eq. 25 is applied, has been changed to a virtual dynamic sensor when direct E_{ISE} readings are converted to a ΔE_{ISE} *vs.* pH plot, *i.e.*, complex formation curves.

VIII. Virtual Potential and Voltammetry as a Virtual Potentiometric Sensor

Similarity in trends observed for the two curves in Figure 9 strongly suggests that, in principle, it should be possible to reproduce the trend in the polarographic corrected shift $CS_{pH(i)}$ (Eq. 18) when changes in potential of an ISE, $\Delta E_{ISE,pH(i)} = E_{ISE}(M) - E_{ISE}(comp)_{pH(i)}$, are plotted *vs.* pH. In fact, the same must apply to any potentiometric metal ion sensor. However, for this to apply some conditions must be met, namely:

1. Polarographic and potentiometric data must be obtained on the same sample solution. This means that all experimental data, such as an initial volume of a sample solution, initial concentrations of a ligand and a metal ion, consecutive volumes of titrant added, pH readings, *etc.* are the same for both experimental techniques.
2. Eq. 18 must be equal to Eq. 25.

The first condition is easily achievable by making use of a dedicated equipment and computer controlled protocols, as described in Section XI. The second condition implies that:

$$CS_{\mathrm{pH}(i)} = E(\mathrm{M}) - E(\mathrm{comp})_{\mathrm{pH}(i)} - \frac{RT}{nF}\ln\left(\frac{I(\mathrm{comp})}{I(\mathrm{M})}\right)_{\mathrm{pH}(i)} = \Delta E_{\mathrm{ISE,pH}(i)} \quad (37)$$

where $\Delta E_{\mathrm{ISE,pH}(i)} = E_{\mathrm{ISE}}(\mathrm{M}) - E_{\mathrm{ISE}}(\mathrm{comp})_{\mathrm{pH}(i)}$. One can imagine several scenarios for the equality $CS_{\mathrm{pH}(i)} = \Delta E_{\mathrm{ISE,pH}(i)}$ to hold, among them:

1. Polarographic signals, recorded at various pH values, must represent a fully reversible electrochemical process. This means that in the case of DPP, the peak for the metal ion solution in the absence of a ligand, $E_{\mathrm{p}}^{\mathrm{r}}(\mathrm{M})$, and a full set of peaks recorded throughout a titration experiment after the ligand addition, the $E_{\mathrm{p}}^{\mathrm{r}}(\mathrm{comp})_{\mathrm{pH}(i)}$ values, must be regarded as fully reversible. Hence, for all practical reasons, they represent fast heterogeneous kinetics and reduction process(es) that are solely controlled by diffusion. A few straightforward protocols that are useful for testing electrochemical reversibility are described in Section XII.
2. From Eq. 25, the potentiometric sensor must respond to the change in the free metal ion concentration as closely as possible to the classical Nernstian slope, as $\Delta E_{\mathrm{ISE,pH}(i)}$ must equal to $(RT/nF)\ln([\mathrm{M}]_{\mathrm{T}}/[\mathrm{M}])_{\mathrm{pH}(i)}$. Clearly, it is necessary to monitor the performance of a potentiometric sensor by monitoring its response slope and some useful protocols are described in Section X.
3. The intensity (or the height) of a polarographic signal, after correction for dilution, does not vary throughout an experiment, *i.e.*, when $I(\mathrm{comp})_{\mathrm{pH}(i)} = I(\mathrm{M})_{\mathrm{pH}(i)}$, then logarithmic term becomes 0 and is no longer a part of Eqs. 18 or 37.

If the above scenarios hold, then $CS_{\mathrm{pH}(i)} = E_{\mathrm{p}}^{\mathrm{r}}(\mathrm{M}) - E_{\mathrm{p}}^{\mathrm{r}}(\mathrm{comp})_{\mathrm{pH}(i)}$ computed from DPP data should be equal to $\Delta E_{\mathrm{ISE,pH}(i)} = E_{\mathrm{ISE}}(\mathrm{M}) - E_{\mathrm{ISE}}(\mathrm{comp})_{\mathrm{x}(i)}$ computed at the same pH(i) values for ISE-generated data.

Regrettably, the requirement (1) of the above scenario seldom holds, meaning that the reduction process is not fully reversible. However, when the departure from electrochemical reversibility is not severe, one can retrieve from experimental curves the $E_{\mathrm{p}}^{\mathrm{r}}$ or $E_{1/2}^{\mathrm{r}}$ values that are expected for a reversible process; protocols suitable for this purpose are described in Section XII. When these protocols fail, one must not use polarographic data for computing formation constants. A similar statement applies to the performance of a potentiometric sensor — if the test in Section X fails then the sensor must be seen as not suitable for rigorous speciation studies.

Typically, the intensity of polarograms decreases more than the expected magnitude predicted from normalization using the applicable dilution factor at a given titration stage. One must understand that many species might be present in a solution at any time during a titration experiment. Hence, it is rather impossible to establish the exact sources of a signal decrease and attribute them quantitatively to each complex formed. Fortunately and regardless of the origin of the observed decrease in I_p or I_d, Eq. 37 takes care of this common occurrence by converting the relative polarographic signal decrease, the $\{I(\text{comp})/I(\text{M})\}_{\text{pH}(i)}$ ratio, to a potential through the $(RT/nF)\ \ln\ \{I(\text{comp})/I(\text{M})\}_{\text{pH}(i)}$ term. As explained in some detail in Section III.B, the latter term predicts by how much the signal should have shifted if its intensity remained the same throughout an experiment. Hence, the $(RT/\text{nF})\ \ln\ \{I(\text{comp})/I(\text{M})\}_{\text{pH}(i)}$ term can be seen as a virtual shift of a reversible polarographic signal, $\Delta E^{\text{r}}(virt)$. Since $I^{\text{r}}(\text{comp})_{\text{pH}(i)}/I^{\text{r}}(\text{M})_{\text{pH}(i)} \leq 1$, the virtual shift is always negative and because it is subtracted from the experimentally observed shift, $\Delta E_{\text{p}}^{\text{r}}(exp) = E_{\text{p}}^{\text{r}}(\text{M}) - E_{\text{p}}^{\text{r}}(\text{comp})_{\text{pH}(i)}$, it increases the experimentally recorded value. This fully explains why the sum of experimental and virtual shifts in a polarographic signal is called the corrected shift,

$$CS_{\text{pH}(i)} = \Delta E_{\text{p}}^{\text{r}}(exp) + \Delta E^{\text{r}}(virt) \tag{38}$$

When a potentiometric sensor is calibrated prior a speciation experiment, a series of sample solutions with known concentrations of a metal ion, but without a ligand, is prepared in a background electrolyte of choice to keep ionic strength and activity coefficients constant. It is then expected that a good potentiometric sensor should respond linearly to a change in the free metal ion concentration on a logarithmic scale with a slope approaching that predicted by the Nernst equation (RT/nF). Typically, an ISE is expected to produce a linear response in the range between 0.1 and 1×10^{-5} M. Notably, when metal ions are not complexed then $[\text{M}] = [\text{M}]_{\text{T}}$.

A speciation experiment with a fixed $[L]_T$: $[M]_T$ concentration ratio is designed such that the initial total metal ion concentration $[M]_T$ is well within the linear response of a given ISE. In addition, the pH of a solution is varied by small-volume additions of a relatively concentrated and standardized OH^- solution. This protocol is employed to avoid significant dilution and keep the total concentration of a metal ion in the linearity range of the ISE. Importantly, if the latter condition is met then the ISE should respond to the free metal ion concentration in a much wider range. An example of the ΔE_{ISE} *vs.* log [M] relationship obtained from the study of Cd(II) complexes with glycine by Cd-ISE is shown in Figure 10.[28] Clearly, one does not know the free metal ion concentrations when metal complexes are formed. The only way to obtain [M] is to solve a full set of mass-balance equations for an assumed M–L–OH model and this also includes incorporation of all known competing complexes, such as $M_p(OH)_r$.

The variation in ΔE of an ISE *vs.* log [M] seen in Figure 10 could hardly be more linear; the regression parameter, R^2 shown to the fourth decimal place, departs from 1 only by 2×10^{-6}. In addition, the slope of 29.08 mV per log [M] unit is very close to the theoretical one. These

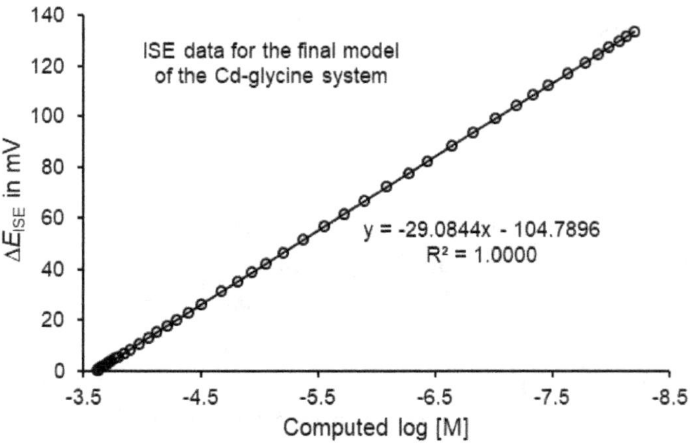

Figure 10. Change in an ISE potential plotted *vs.* the free metal ion concentration computed by solving mass-balance equations for the final M–L–OH model with refined formation constants. $[L]_T$:$[M]_T$ = 100.7. Initial $[M]_T$ = 2.4947×10^{-4} M. Adapted from Ref. 28.

observations strongly support the final model that incorporates M(HL), ML, ML_2, ML_3, $ML_2(OH)$ and all known $M_p(OH)_r$ complexes where M = Cd(II) and L = a fully deprotonated glycine, *i.e.*, the glycinate anion. Meeting the $CS_{pH(i)} = \Delta E_{ISE,pH(i)}$ equality implies that voltammetry (or polarography) can be treated as a virtual potentiometric sensor. If this is the case, then polarography must produce linear $CS_{pH(i)}$ *vs*. log [M] relationships of a comparable quality as seen in Figure 10 for ISE-generated data. The Cd(II)–glycine system, for which a nearly perfect linear trend was obtained from the ISE studies (Figure 10), was used to test the performance of polarography as a virtual potentiometric sensor.

Data obtained from a DPP experiment ran at the $[L]_T$:$[M]_T$ concentration ratio of 400 (see Figures 7–9) were used for this purpose. Figure 11 shows experimental (circles) and computed (pluses) corrected shifts that are plotted *vs*. log [M]; trend lines fitting the computed $CS_{pH(i)}$ are shown as dotted lines. It is necessary to state that both the computed $CS_{pH(i)}$ and $[M]_{pH(i)}$ values were obtained using MBEs solved for a specific M–L–OH model. Data obtained for three models are shown in Figure 11:

1. Part (a) was obtained for major complexes ML, ML_2, and ML_3 identified in Sections V.A and V.B using initial estimates from Table 3 in Section VI: log β_{110} = 4.49, log β_{120} = 7.71, and log β_{130} = 9.91. As was the case with the solid line CCFC-pH in Figure 8, Section VII.B, the pluses in Figure 11a are above the experimental corrected shifts and this is due to the initial estimates of the formation constants being too large. Interestingly, the middle part of the circles follows the trend of the dotted line quite well with a response slope quite close to that predicted from a Nernst equation. Curvature is observed at both ends of the experimental (circles) and computed (pluses) corrected shifts. This clearly shows that additional complexes are required and one can arrive at the same conclusion by considering the large departures of the computed corrected shifts from the dotted trend line.
2. Part (b) represents the same model with ML, ML_2, and ML_3 complexes but with refined formation constants: log β_{110} = 4.26, log β_{120} = 7.17, and log β_{130} = 9.80. Much improved linearity is observed for both, experimental and calculated, corrected shifts. Importantly, as also seen for the dashed line CCFC-pH in Figure 8, Section VII.B, a

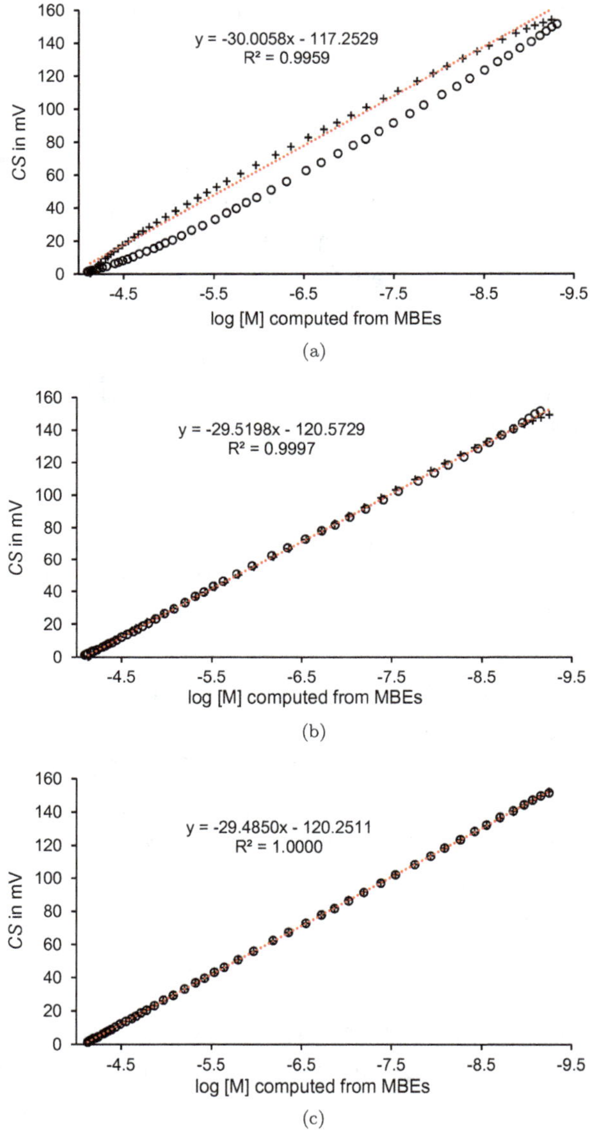

Figure 11. Experimental (circles) and computed (pluses) corrected shifts plotted against the free metal ion concentrations obtained from MBEs. The dotted line represents a trend line obtained for the computed *CS* values (pluses). The computed *CS* and log [M] values were obtained by solving MBEs for models containing all $M_p(OH)_r$ complexes plus part (a) — ML, ML_2, and ML_3 with initial estimates of formation constants (Section VI); part (b) — ML, ML_2, and ML_3 with refined initial estimates of formation constants (Figure 8 in Section VII.B); part (c) — the final model containing M(HL), ML, ML_2, ML_3, and $ML_2(OH)$ with refined formation constants. Unpublished data except experimental points coming from Ref. 28.

clear departure from linearity is observed for the largest and smallest log [M] values. This fully confirms the presence of additional complexes as identified from Figure 11a.

3. Part (c) shows how exceptionally well the DPP experiment performed as a virtual potentiometric sensor. The experimental and computed trends are nearly ideal across a very large free metal ion concentration range. The final model includes (i) M(HL) that was added at the largest log [M] values, *i.e.*, at the lowest pH values as this complex is expected to start forming prior the ML_q complexes, (ii) ML, ML_2, ML_3 predicted by protocols described in Section V, as well as (iii) $ML_2(OH)$ that was added at the lowest log [M] values, *i.e.*, at the highest pH values where hydrolyzed complexes are expected to form. The exceptional linear trend and superimposition of experimental and computed points was obtained with the following refined formation constants: log β_{111} = 10.07, log β_{110} = 4.20, log β_{120} = 7.30, log β_{130} = 9.64, and log β_{121} = 11.49. The same values produced an exceptional fit of the CCFC-pH into the ECFC-pH in Figure 9, Section VII.B.

ESTA[47–50] (Equilibrium Simulation by Titration Analysis) is a powerful suite of software modules dedicated to the potentiometric data in the study of metal-ligand equilibria. Its robust modules can be used to refine data obtained from measurements of the free proton [H] or free metal ion [M] concentrations. Glass electrode potentiometry (GEP) is the most powerful and versatile analytical technique in the study of equilibria and the majority of known formation constants and protonation constants were determined by use of GEP. Not surprisingly, excellent dedicated software, such as ESTA, was developed for that purpose. The added advantage of ESTA is the ability to refine many additional parameters during refinement operations employed to compute formation constants. Moreover, one can run a single refinement operation simultaneously on several sets of data coming from different potentiometric techniques.

It is obvious that it would be a great advantage in modelling M–L–OH equilibria if polarographic data could also be submitted to ESTA. The ultimate goal would be to refine polarographic and potentiometric data coming from either a single experiment or different experiments.

To achieve that one must provide ESTA with readings of glass electrode potentials (or potentials obtained using a metal-specific potentiometric sensor) that were recorded throughout a titration of a sample solution containing M and L.

Realizing that ESTA (or any other powerful software dedicated to potentiometry) cannot distinguish between true and virtual potentials, it should be possible to make use of polarographic (non-equilibrium and dynamic) data. To achieve that one must define $E^{\mathrm{r}}(virt)$ using experimental polarographic data,

$$E^{\mathrm{r}}(virt)_{\mathrm{pH}(i)} = E^{\mathrm{r}}(\mathrm{comp})_{\mathrm{pH}(i)} + \frac{RT}{nF}\ln\left(\frac{I(\mathrm{comp})}{I(\mathrm{M})}\right)_{\mathrm{pH}(i)} \tag{39}$$

$$E^{\mathrm{r}}(virt)_{\mathrm{pH}(i)} = E^{\mathrm{r}}(\mathrm{comp})_{\mathrm{pH}(i)} + \Delta E^{\mathrm{r}}(virt)_{\mathrm{pH}(i)} \tag{40}$$

Figure 12 shows that indeed virtual potentials follow the Nernst equation nearly exactly, *i.e.*, the slope obtained (29.49 mV/log-unit) is only 0.09 mV smaller when compared with a theoretical value for the reversible reduction of Cd^{2+} ions at 25 °C — virtual potentials were computed for DPP data shown in Figure 11c.

Finally, for ESTA to make use of potentiometric data, one must also provide parameters obtained from the calibration of the potentiometric sensors as:

$$y = E' + slope \times \log\,[(\mathrm{H,M})] \tag{41}$$

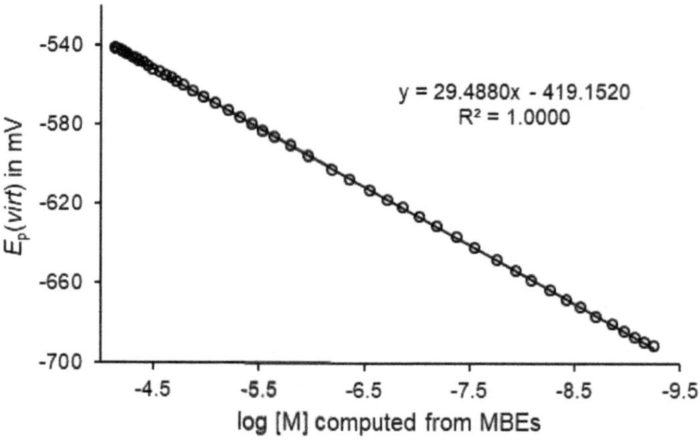

Figure 12. DPP virtual potentials plotted *vs.* the free metal ion concentrations computed from MBEs solved for the M–L–OH system described in Figure 11c. Unpublished data — experimental data from Ref. 28 were used.

where y stands for the reading of potential. Usually, the formal potential E' and the response *slope* come from fitting calibration data recorded (i) during an acid–base titration with a glass electrode ([H] is exactly known at each titration step) or (ii) on a series of solutions with a free metal ion with exactly known $[M] = [M]_T$ values.

It is obvious that one cannot calibrate polarographic data in the same way. This is because the E_p and $E_{1/2}$ values do not depend on the concentration of a free metal ion when no ligand is present; only intensities of signals, I_p and I_d, vary and they increase linearly with an increase in $[M]_T$ in an entire concentration range available to the selected mode of polarographic technique. Hence, to employ ESTA for the refinement of virtual polarographic data in the form of virtual potentials (Eqs. 39 and 40) the response *slope* in Eq. 41 is set to the ideal Nernst value for a given temperature and number of electrons n, *e.g.*, for cadmium(II) ions $n = 2$. In the case of the formal potential E', a good guess must be provided as an input to ESTA and it has to be refined together with formation constants — detailed protocols are reported elsewhere.[32]

IX. Species Distribution Diagrams in Designing Experiments and Verifying the Proposed M–L–OH Model

The species distribution diagram (SDD), *i.e.*, a plot showing variation of %-fraction of M- and L-containing species *vs.* pH (or pL), is an excellent visual and easy to interpret tool used to predict the best experimental conditions. This is because one can explore the impact of all species and each formation and protonation constant as well as the total ligand and metal ion concentrations on the solution composition in the entire pH or pL range. Additionally and of paramount importance, one must incorporate the best available solubility product (preferably determined for freshly precipitated metal hydroxides) for the assumed experimental conditions, such as background electrolyte, ionic strength and temperature when computing the SDDs. Actually, when experimental values are not available, it is recommended to include best estimate in K_{sp} and design an experiment such that precipitation is predicted at as high pH as possible. In any case, one should always test the predictions made prior running full titration experiment.

Having established an initial model with preliminary formation constants, either based on the literature data, or applying linear free energy relationships[13] to carefully selected ligands and metal ions, or simply by doing an initial experiment, one can design experimental conditions needed to obtain the best possible experimental results. In general, an acid–base titration gives a better overall perspective of and insight into the metal-ligand system and hence this should be considered first. It is preferred and is a good common practice to study the system starting from low pH to assure the smallest degree of complexation and minimize hydrolysis. A gradual increase in pH promotes complex formation reactions by increasing the free ligand concentration due to the deprotonation of the ligand progressing with the pH increase. Moreover, with progressive complex formation the free metal ion concentration decreases. Because of that, precipitation of metal hydroxo species and formation of polynuclear hydroxo-complexes can be minimized or avoided altogether.

A. Species Distribution Diagrams as a Function of pH

For the purpose of exploring the applications and to illustrate usefulness of the SDDs, two Cd^{II}-L-OH systems will be modeled. The same sets of all relevant Cd(II) formation constants (including hydroxo-complexes), protonation constants of the ligands glycine and picolinic acid were used as reported in original papers[28,43] — for convenience they are included in Table 4.

The experiments were performed at $I = 0.5$ (H,Na)NO_3 M and at 25 °C. Ideally, one would like to incorporate all necessary data that are directly applicable to the experimental conditions envisaged but often this is not achievable. Fortunately, for preliminary modelling purposes, most suitable and highest quality data are not always necessary and this is particularly true when strong metal complexes are expected to form. Just as an example, the difference in log β values for Cd(II) complexes with picolinic acid at ionic strength of 0.1 and 0.5 M are on the second decimal place[43] and this will have no impact on predicted species distribution diagrams whatsoever under typical experimental conditions used in polarographic studies.

Table 4. Solubility product K_{sp} and overall formation constants for Cd(II) complexes with OH^-, picolinic acid and glycine, dissociation constant for water, stepwise protonation constants for the ligands picolinic acid and glycine,[11,12] used in computing species distribution diagrams and expected shifts in potential, *E*-shifts, shown in Figures 13–21.

Ligand	Equilibrium	log β	*I*/M
OH^-	$H_2O \rightleftarrows H^+ + OH^-$	−13.74	0.5
	$Cd^{2+} + OH^- \rightleftarrows [Cd(OH)]^+$	4.0	3.0
	$Cd^{2+} + 2OH^- \rightleftarrows [Cd(OH)_2]$	7.7	3.0
	$Cd^{2+} + 3OH^- \rightleftarrows [Cd(OH)_3]^-$	10.3	3.0
	$Cd^{2+} + 4OH^- \rightleftarrows [Cd(OH)_4]^{2-}$	12.0	3.0
	$2Cd^{2+} + OH^- \rightleftarrows [Cd_2(OH)]^{3+}$	5.06	3.0
	$4Cd^{2+} + 4OH^- \rightleftarrows [Cd_4(OH)_4]^{4+}$	24.9	3.0
	$Cd(OH)_2(s) \rightleftarrows Cd^{2+} + 2OH^-$	−14.3	3.0
Picolinic acid	$M + L \rightleftarrows ML$	4.29	0.5
	$M + 2L \rightleftarrows ML_2$	7.89	0.5
	$M + 3L \rightleftarrows ML_3$	10.49	0.5
Glycine	$M + L \rightleftarrows ML$	4.18	0.5
	$M + 2L \rightleftarrows ML_2$	7.50	0.5
	$M + 3L \rightleftarrows ML_3$	9.76	0.5
		log K^H	
Picolinic acid	$L^- + H^+ \rightleftarrows HL$	5.18	0.5
	$HL + H^+ \rightleftarrows H_2L^+$	0.86	0.5
Glycine	$L^- + H^+ \rightleftarrows HL$	9.54	0.5
	$HL + H^+ \rightleftarrows H_2L^+$	2.39	0.5

The first system involves the ligand picolinic acid and looking at its protonation constants it is obvious that it will be singly deprotonated already at very low pH values. Moreover, due to the presence of the electron donating pyridine ring, it is expected that picolinic acid will start forming complexes with Cd(II) well below pH 2, a pH-range not best suited for GEP studies. It is very useful and indeed highly recommended to generate a number of SDDs that might be suitable for different analytical techniques and as such, we will focus on polarography, GEP and ISE. Importantly, it should be recognized that very different parameters are measured and utilized by these techniques and this will also determine the optimal conditions for the dedicated experiment. All known (or assumed) constants must be kept fixed during solving mass balance equations when SDDs are computed. Hence, there are three experimental variables one can "play" with, namely $[M]_T$ and $[L]_T$, as well as their ratio.

Polarography works well in a wide range of the total metal ion concentration (between 10^{-2} and 10^{-6} M) but it requires a significant excess of a ligand when labile systems are studied, typically $[L]_T:[M]_T \geq 30$. One must also consider the solubility of organic ligands and their possible negative impact on the quality of recorded polarograms due to (i) adsorption of a ligand on the surface of the working mercury electrode as well as (ii) difficulty in maintaining constancy in ionic strength of a sample solution, when ligand is present at very high concentrations. It is then reasonable to start modelling with $[M]_T = 1 \times 10^{-4}$ M and $[L]_T:[M]_T = 200$ to promote formation of complexes with moderate $[L]_T = 2 \times 10^{-2}$ M. The SDD shown in Figure 13a shows several important features:

1. A large pH window is available for polarographic studies, up to pH of ~11.6. Above this pH precipitation of $Cd(OH)_2(s)$ is predicted as indicated by a vertical dashed-line.
2. If no other complexes are formed, except ML, ML_2, and ML_3, it is sufficient to execute the experiment starting from pH of about 1.5 (if possible) and record the final polarogram at pH of ~6 as the solution composition does not vary significantly at pH > 6.
3. In general, it is best to avoid the region below pH 2 (as is discussed in Section IX.E). This means that insufficient initial data (1< pH < 2) will be recorded and this, in principle, might have some impact on the accuracy in the refined log β_{110} for the ML complex.
4. To establish whether $CdL_q(OH)_r$ complexes are formed at all, one should initially record polarograms in pH-steps of 1 up to pH of 11; precipitation should be avoided as it might negatively impact the

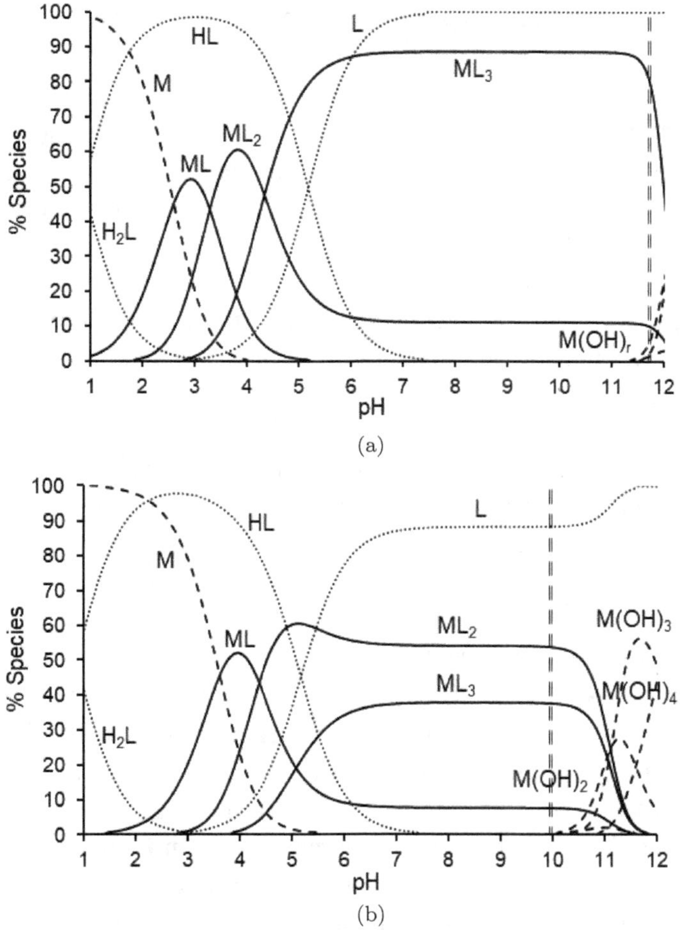

Figure 13. SDDs computed for the Cd(II)-picolinic acid system to be studied at $[M]_T = 1 \times 10^{-4}$ M by: (a) polarography with $[L]_T = 2 \times 10^{-2}$ M giving $[L]_T{:}[M]_T = 200$; (b) ISE with $[L]_T = 2 \times 10^{-3}$ M giving $[L]_T{:}[M]_T = 20$. The vertical dashed-line shows the pH at which $Cd(OH)_2$(s) precipitation is predicted. Unpublished data.

performance of a reference and working electrode. Due to constancy in the solution composition, no shift of the recorded polarogram along the applied potential scale should be observed between pH 7 and 11 in the absence of $CdL_q(OH)_r$ complexes. In other words, if a polarographic signal shifts above pH 6 then it strongly suggests formation of hydrolyzed complexes.

It is then clear that one particular set of conditions may not be conducive to providing the most accurate log β values for all species. Recalling that the smaller excess of a ligand the more delayed complex formation reactions are, a SDD for the $[L]_T:[M]_T = 20$ was computed with an aim of avoiding collecting data at pH < 2 — see Figure 13b.

A 10-fold decrease in the total ligand concentration reveals that for the same $[M]_T = 1 \times 10^{-4}$ M:

1. The pH-window available for collecting data decreased as precipitation is now predicted at pH = 10. The indicated %-fractions for $Cd_p(OH)_r$ complexes must be seen as hypothetical — the solution composition at pH > 10 is applicable only if precipitation does not occur.
2. Indeed and importantly, formation of complexes is postponed and collecting data from pH of about 2 onwards should provide sufficient data-points for computing the formation constant for ML.
3. The quality of log β_{130} might be somewhat questionable, as ML_3 never becomes a major species in a solution.
4. The assumed experimental conditions are not perfect for polarographic studies of labile systems as $[L]_T:[M]_T$ of 20 is or might be a bit too small but one might consider making use of a Cd-ISE. However, one can employ polarography (or combined DPP/ISE experiment) by increasing the ligand to metal ratio to 30–40 as this will not change formation of complexes seen in Figure 13b dramatically when ML is considered but might improve conditions for the ML_3 formation significantly. In any case, it is recommended to recalculate SDDs for analytical conditions predicted to be the best for a purpose.

For both $[L]_T:[M]_T$ ratios discussed above, the ligand composition is similar at lower pH values; it is seen that as pH increases and more ligand is found in the singly protonated HL form, formation of ML_q species is facilitated. It should be noted, however, that it is not only the $[L]_T:[M]_T$ ratio that is important, but the actual concentration of the total metal ion in solution plays an important role too. This is illustrated in Figure 14 by the SDD where a classical polarographic $[L]_T:[M]_T = 200$ condition was used, but both $[M]_T$ and $[L]_T$ are an order of magnitude lower than that in

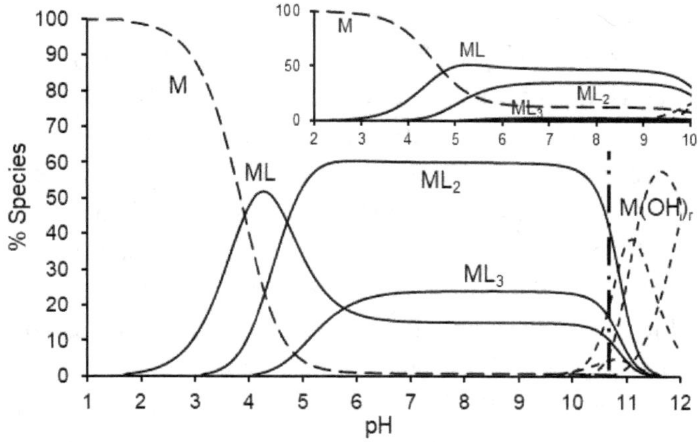

Figure 14. SDD for the Cd(II)-picolinic acid system. $[M]_T = 1 \times 10^{-5}$ M and $[L]_T = 2 \times 10^{-3}$ M giving $[L]_T{:}[M]_T = 200$. For comparison, the inset shows the SDD where $[M]_T = 1 \times 10^{-5}$ M and $[L]_T{:}[M]_T = 20$. Unpublished data.

Figure 13a. In this case, even with the high $[L]_T{:}[M]_T$ ratio, the overall lower concentrations result in complex formation driven to a lesser extent. Although this is not a perfect scenario, it is seen in Figure 14 that collecting many experimental points in the pH range between 2 and 6 should provide excellent estimates for ML and ML_2 formation constants. Ideally, to produce statistically reinforced log β values, a few titration experiments should be performed under significantly different conditions and, if possible, using different analytical techniques. Clearly, by controlling the solution conditions one can bias the various species formed across the pH range, including postponing precipitation or avoiding the need to work at low pH values. The inset in Figure 14, with $[L]_T{:}[M]_T = 20$ and $[M]_T = 1 \times 10^{-5}$ M, shows that these experimental conditions are the least recommended, if at all. Almost no ML_3 is formed and %-fractions of ML and ML_2 are small and none becomes a dominant complex. It is seen that complexes start to form only above pH of 3 and one might be tempted to consider this as an excellent experimental set-up for determining formation constant of ML. However, the shifts in potential that would be measured under these conditions must still be large enough and these values can be estimated using the tool described in Section IX.B. Data presented in Figure 14 must be seen as a border-line case study of a general nature as conditions employed are neither suitable for ISE (too low $[M]_T$) nor polarography as $[L]_T$:$[M]_T$ = 20 is too low for fully labile systems.

Nonetheless, data in Figures 13 and 14 are perfectly suited for predicting the best conditions for either DCP, DPP or ISE. Note that in all the SDDs presented thus far, the only hydrolysis products seen were $[Cd(OH)_r]^{(2-r)+}$, where $r = 1$–4, even though $[Cd_2(OH)]^{3+}$ and $[Cd_4(OH)_4]^{4+}$ were included in the M–L–OH model. This is an important observation as, generally, at the low total metal concentrations, formation of these or any other possible polynuclear species is not favorable and often can be avoided altogether making speciation studies of new systems much easier.

When employing GEP it is solely the free proton concentration that is monitored and the change in proton concentration can be ascribed to at least six processes taking place throughout an acid-base titration, namely:

1. The addition of a strong standardized base to a sample solution.
2. The deprotonation of the ligand under the changing pH conditions that can be quantified based on the known protonation constants.
3. The deprotonation of the ligand due to the release of protons on complex formation with the metal ion.
4. The release of protons from water when metal hydroxo complexes are formed.
5. A change in proton concentration due to dissociation of water.
6. An unwanted presence of CO_2 (from the atmosphere) that will react with water to produce carbonic acid and, in some cases, form metal carbonato complexes.

This clearly shows how much attention and care has to be taken to produce reliable GEP data — simply speaking, this is a highly demanding analytical technique where experimental conditions and execution of an experiment are concerned. Thus, the formation of metal complexes with a ligand L should be the main contributor of proton's concentration in order to provide accurate log β values. To achieve that, relatively large concentrations of both $[M]_T$ and $[L]_T$ are required, with low $[L]_T$:$[M]_T$ ratios, being close to that formed by the complexes (*i.e.* $[L]_T$:$[M]_T$ = 1–3 in this case).

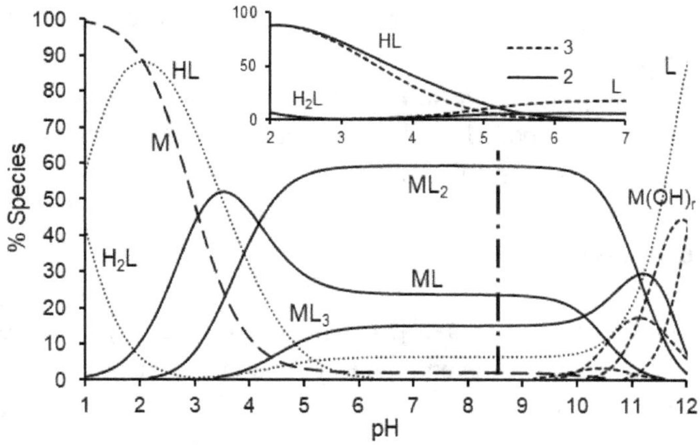

Figure 15. SDD for the Cd(II)-picolinic acid system. It was computed for experimental conditions suitable for GEP as well as ISE. $[L]_T:[M]_T = 2$ and $[M]_T = 5 \times 10^{-3}$ M. The inset compares the SDD for the ligand species for $[L]_T:[M]_T = 2$ and 3 (solid and dashed lines, respectively) and $[M]_T = 5 \times 10^{-3}$ M for both ratios. Unpublished data.

Figure 15 shows the SDDs for typical GEP and ISE conditions having $[L]_T:[M]_T = 2$. It is clearly seen that the %-fraction of protonated ligand species change dramatically along the pH axis mainly due to formation of Cd(II)-picolinic acid complexes rather than the addition of a base solution, as exemplified by very low concentrations (%-fractions) of all forms of the ligand when complexes of Cd(II) are formed, *i.e.*, for pH > 5.

Although a high $[M]_T$ is usually preferred in order to provide a larger change in proton concentration, it is clear that formation of the ML species starts below pH 2 already (Figure 15). GEP is even more susceptible to errors in the pH range below 2. Hence, slightly reducing $[M]_T$ would shift formation of complexes to higher pH values and provide better experimental conditions for this particular M–L–OH system. Also, only when the $[L]_T:[M]_T$ ratio is increased to 3 will the ML_3 complex also be significantly formed. It would thus be recommended that experiments be run at various $[L]_T:[M]_T$ ratios, where a ratio of 1 will provide the most accurate formation constant for the ML and hence further experiments at higher ratios are required to promote the formation of ML_2 and ML_3.

Unfortunately, these successive formation constants also become less reliable. The inset in Figure 15 highlights the small difference in the %-fraction of HL and L for experiments with $[L]_T$:$[M]_T$ ratios of 2 and 3, represented for the same $[M]_T$. This indicates the need for high concentrations of reactants and careful experimentation.

Utilizing ISEs in complex formation studies is probably the most versatile in terms of:

1. The concentration ranges of $[M]_T$ that can be used, typically between 0.1 and 1×10^{-4} M or even lower; it depends on the kind of electrode used and its performance. Importantly, one must not work at $[M]_T$ close to the limit of linearity in the ISE response; in general, it is best to start at fairly high $[M]_T$ as it promotes formation of complexes.
2. The ISE linear response to the free metal ion that can span over several orders of magnitude more when compared to the $[M]_T$ range.
3. A large range of $[L]_T$:$[M]_T$ ratios that can vary from about 1:1 as for GEP to 1000:1, as in polarography. It is important to make sure that the ligand, particularly at high concentrations, does not affect the electrode performance due to interactions with the membrane.
4. Data collected can be submitted to software applications dedicated to either GEP or polarographic studies.

The fact that these electrodes are not used more in these types of studies stems from them being less stable than the other methodologies considered. Once again, the pH of the solution is also measured in this case to enable all the required mass balance equations to be solved.

B. Predicting Shifts in Potential, ΔE

Another extremely important tool to consider when planning polarographic/ISE experiments is to estimate the shifts in the recorded polarograms/ISE-response (ΔE) that would be measured for a particular set of experimental conditions. It is obvious that the larger the change in potential, the better the accuracy of the determined formation constants. It is possible to predict a trend in ΔE values as a function of pH (or pL) from Eq. 17 and one must use just the right-hand side for this purpose, Eq. 42:

$$\Delta E = \frac{RT}{nF}\ln\left(\frac{[\mathrm{M}]_\mathrm{T}}{[\mathrm{M}]}\right)_{\mathrm{x}(i)} \tag{42}$$

Clearly, Eq. 42 is equally applicable to any mode of polarography (DC or DP) regardless of whether the metal-ligand system is reversible or not, labile or non-labile. Eq. 42 is also equally applicable to an ISE with Nernstian response; in case of non-Nernstian response of ISE, one must use Eq. 25, *i.e.*, replace the *RT/nF* term by the experimentally determined *slope* obtained for a specific ISE. Actually, predicting the trend ΔE *vs.* pH (pL) is exactly the same mathematical operation as generating the CCFC; both operations require the full M–L–OH model and the full set of formation and protonation constants to compute the free metal ion concentration at a specific x(i) value. A typical use of CCFCs for modelling of a M–L–OH system and refinement of formation constants described in Section VII involved variation in the formation constants in order to reproduce experimental points as accurately as possible. When predicting experimental conditions, one does not have experimental points, but can assume a model and formation constants for which Eq. 42 can be computed, either as a single point calculation or a CCFC in the entire pH/pL range.

A comparison of the *E*-shifts (or ΔE values) calculated at various pH values for the conditions used for computing the above SDDs (shown in Figures 13 and 14) is presented in Figures 16a and 16b, respectively. These indicate that (i) the higher the ligand and metal ion concentrations (Figure 16a) and (ii) the larger $[\mathrm{L}]_\mathrm{T}$:$[\mathrm{M}]_\mathrm{T}$ ratio (open circles in Figure 16), the larger the *E*-shifts predicted. Moreover, the largest shifts are observed in the regions where the degree of complex formation is largest. When a plateau is observed in Figure 16, *i.e.*, where the maximum shift in potential was attained, the composition of solutions remain virtually unchanged as seen in Figures 13 and 14.

Considering the Cd(II)-picolinic acid system, it is now clear that proceeding with the pH titration beyond pH ~7 under the assumed conditions and model will not provide additional information. However, that being said, it would depend on how good the initial estimates are regarding the model and formation/protonation constants. If these estimates were not based on initial experimental data but rather on extrapolated information

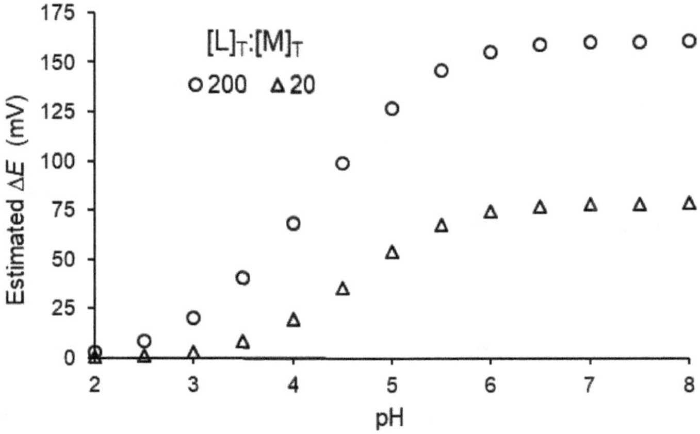

(a)

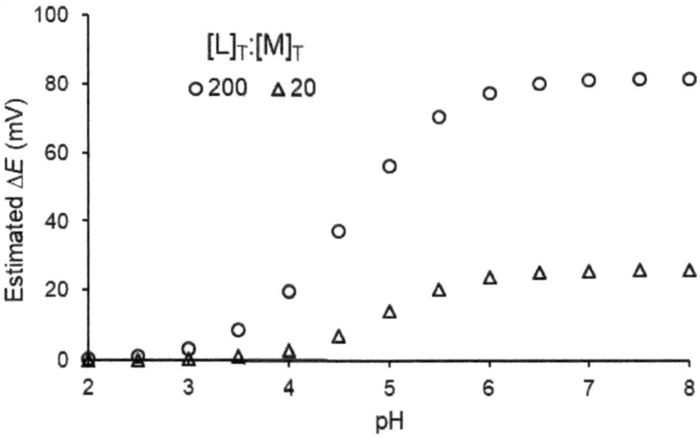

(b)

Figure 16. Estimated potential shifts ΔE for the Cd(II)-picolinic acid system where (a) $[M]_T = 1 \times 10^{-4}$ M and (b) $[M]_T = 1 \times 10^{-5}$ M at the indicated $[L]_T$:$[M]_T$ ratios. Unpublished data.

from other systems, it is highly recommended to extend the titration to higher pHs for the first exploratory experiment. If an additional shift in E was recorded where the plateau was predicted in Figure 16, then formation of new complexes would be positively identified.

Of particular note is the low pH region. As shown in Figure 13, the ML complex starts forming below pH 2, but no significant shift in potential is predicted at pH 2 — see open circle points in Figure 16a. Hence, concerns related to collecting data at pH < 2 expressed above are no longer valid and it would be acceptable to start measurements from pH = 2 without compromising the significance of acquired data points and quality of determined formation constants. This emphasizes the strength of having multiple predictive tools, which can be used in designing experiments.

C. Species Distribution Diagrams as a Function of the Total Ligand Concentration

One might attempt to improve formation constants obtained from acid–base titrations by performing complementary ligand titrations. In this case, a carefully selected pH is kept constant while increasing the ligand concentration to promote complex formation. Information obtained from the acid-base titrations is used as a basis on which to select the pH at which ligand titration experiments should be performed. Typically, a concentrated ligand titrant solution should be prepared; this requirement, however, might not be met for ligands with limited solubility at specifically selected pH values. Finally, only complexes identified or predicted at a given pH would be formed, so this approach, *i.e.*, the ligand titration, cannot provide the full overview of all species, but allows one to focus on one or two species at a time.

From data shown in Figure 13, where SDDs were computed for $[L]_T$:$[M]_T$ ratios of 200 and 20, it is immediately apparent that the ligand titration experiments should be performed at pH values between 2 and 5. For the lower ligand to metal ratio, ML and ML_3 start to form at pH 2 and 4, respectively. At the larger ratio, these complexes are nearly fully formed at pH values of 3 (ML) and 5 (ML_3). Moreover, one must note that in this pH-window (2 < pH < 5), the ligand picolinic acid is predominantly in its neutral HL form. This means that increasing the ligand concentration will not have an immense impact on the ionic strength of a solution sample; nonetheless, one must remember that formation constants do depend on ionic strength so much attention must be paid in this regard in planning the ligand titration experiments. Also, as this is purely a computational exercise, it is reasonable to compute SDDs for a wide range of the total ligand concentrations as this should guide one in terms of how to run experiments.

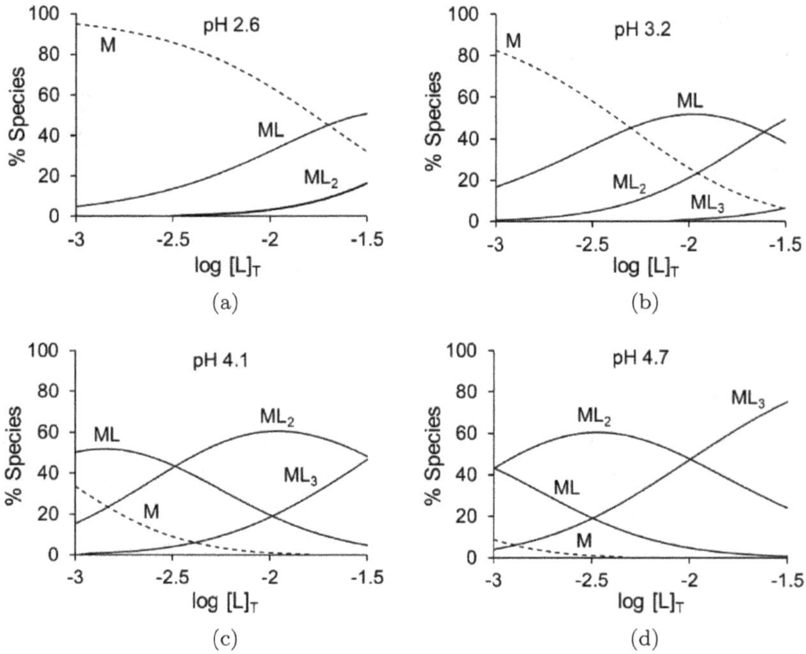

Figure 17. Expected SDDs from ligand titrations of the Cd(II)-picolinic acid system at the indicated pH values (a–d) using $[M]_T = 1 \times 10^{-4}$ M and relevant data in Table 5. Unpublished data.

Figure 17 presents SDDs for the Cd(II)-picolinic acid system at four selected pH values. In each case, the $[L]_T$:$[M]_T$ ratio is increased from 10 to ~316 and $[M]_T$ is constant at 1×10^{-4} M. A rather low value of $[M]_T$ was selected but still the 320 times excess of ligand results in $[L]_T = 3.2 \times 10^{-2}$ M, meaning that one would have to work at high ionic strength. Working with lower $[M]_T$ values of say 1×10^{-5} M, essentially eliminates the use of DCP. Starting from the lowest pH = 2.6 considered, only ML is formed appreciably, with ML_2 always remaining a minor species. Although one might be tempted to run an experiment at this pH, data shown in Figure 18 (diamonds) makes it clear that it would be a fruitless attempt as extremely small *E*-shifts are predicted. Note that at $[L]_T = 1 \times 10^{-2}$ M an *E*-shift of just a few mV is predicted that can be attributed solely to the 30% presence of ML with the remaining 70% of $[M]_T$ being mainly the free metal ion. At pH 3.2, a significant increase in both ML and

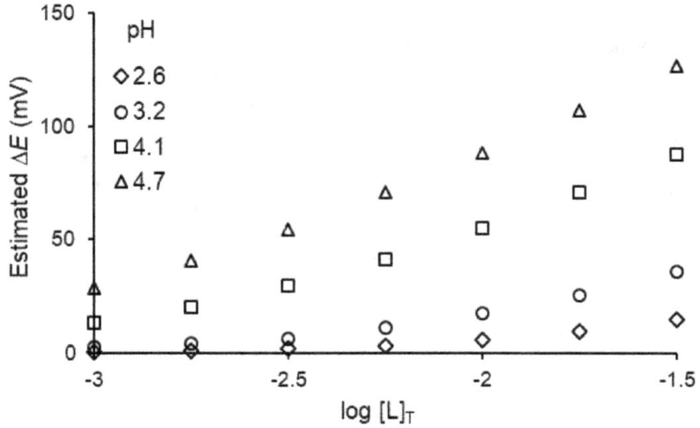

Figure 18. The estimated *E*-shifts predicted for the SDDs shown in Figure 17. Unpublished data.

ML_2 concentrations is seen; only a small amount of ML_3 is formed. Hence, *E*-shifts reaching 30 mV (circles in Figure 18) should provide reasonable estimates for both ML and ML_2 species. At pH 4.1, the concentration of ML is decreasing and hence its contribution to *E*-shifts in the entire range in $[L]_T$ in Figure 17c will provide no useful information regarding this species. Both ML_2 and ML_3 are formed, and even though ML_3 is never a dominant species, significant shifts seen as squares in Figure 18 would provide valuable information about both these species. The SDD at pH 4.7 (Figure 17d) predicts that:

1. [ML] decays to negligible concentrations meaning that essentially it will not contribute to computed *E*-shifts significantly — see triangles in Figure 18.
2. The ML_2 complex becomes a minor species for $[L]_T > 0.01$ M and its contribution to *E*-shifts diminishes starting from $\log [L]_T = -2.5$, and
3. Large shifts seen as triangles in Figure 18 are mainly due to formation of ML_3.

These observations strongly predict that the most accurate estimate in the ML_3 formation constant should be obtained at this pH. This might be crucial information in the case of $\log \beta_{130}$ values, obtained from titrations

Table 5. Estimated ΔE values (E-shifts) calculated for a ligand titration at pH 4.7 based on the theoretical case where $[M]_T$ is fixed, or for practical applications where a titrant (of either 0.1 or 0.01 M) is used and dilution of the metal ion occurs. The volume of ligand required is determined by the particular $[L]_T$:$[M]_T$ ratio. Unpublished data.

	Fixed $[M]_T$		0.1 M Ligand Titrant		0.01 M Ligand Titrant	
$[L]_T$:$[M]_T$	$[M]_T/10^{-4}$ M	E-shift/mV	$[M]_T/10^{-4}$ M	E-shift/mV	$[M]_T/10^{-4}$ M	E-shift/mV
10.0	1.00	28.36	0.990	28.20	0.909	26.82
17.8	1.00	40.44	0.973	39.87	0.783	35.53
31.6	1.00	54.60	0.944	53.11	0.627	43.74
56.2	1.00	70.54	0.896	67.51	0.464	50.60
100.0	1.00	88.05	0.823	82.10	0.317	55.74
177.8	1.00	106.92	0.718	96.07	0.203	59.26
316.2	1.00	126.91	0.585	108.47	0.124	61.49

at pH 4.1 and 4.7, being significantly different. In such a case, one might consider fixing the log β_{130} value, obtained at pH = 4.7, during refinement operations of data collected at lower pH values.

The above analysis suggests that one should consider running experiments at pH values of 3.2, 4.1 and 4.7. Having said that one must realize that SDDs in Figure 17 were computed for a fixed value of $[M]_T$. This, however, is not possible practically as a titrant containing the ligand is added to the sample solution. Table 5 shows the extent to which the metal ion is diluted when using a titrant containing either 0.1 or 0.01 M ligand; it was assumed for each case that $[L]_T$:$[M]_T$ was kept constant and hence the $[M]_T$ would also decrease with increasing dilution.

The estimated ΔE values (E-shifts) in Table 5 were calculated at pH 4.7 based on the actual concentration of $[M]_T$ and $[L]_T$ in solution. Clearly, the 0.01 M titrant is not suitable as:

1. It significantly dilutes the metal ion; note that to reach the $[L]_T$:$[M]_T$ ratio of about 320, the volume of a sample solution would have to increase nearly tenfold as $[M]_T$ decreased from 1×10^{-4} to 1.2×10^{-5} M.
2. Essentially, one cannot practically reach large enough $[L]_T$:$[M]_T$ ratios due to a large increase in sample volume.

3. The value of $[M]_T$ becomes too small for recording DC polarograms with sufficiently large reduction currents. One approaches a detection limit when the total concentration of M drops down to 10^{-5} M and the ratio of reduction to capacitance (background) current becomes unfavorable.
4. Appreciably smaller *E*-shifts would be observed at large ligand to metal concentration ratios, thereby producing much lower quality data in terms of accuracy of the refined formation constants.

Therefore, if the ligand is not soluble enough to generate titrants of at least 0.1 M, a ligand titration should not be considered. Furthermore, one must stress that maintaining I = const, a prerequisite for rigorous speciation studies, might be a challenge when a ligand titration is performed. It is then of importance to report exact initial and final solution composition, pH of a sample solution and dominant forms of charged species of ligand, with a special attention paid to ionic strength constancy (or variation) throughout an experiment.

D. Confirming Additional Species

The second M–L–OH system that will be looked at is the Cd(II)–glycine system — see Figure 19. Relevant data used to compute SDDs are included in Table 4. The SDD shown in Figure 19a was computed for the M–L–OH model containing all $Cd_p(OH)_r$ known species (Table 4) and ML, ML_2, and ML_3 Cd(II) complexes reported in the database of critically selected stability constants of metal complexes compiled by Martell and Smith[11,12] using $[M]_T = 1 \times 10^{-4}$ M and $[L]_T$:$[M]_T = 200$, *i.e.*, typical DCP/DPP conditions. Studies of the Cd(II)–glycine system by polarography and a cadmium ISE suggested the presence of two additional species, M(HL) and $ML_2(OH)$ with log β value around 10.0 and 11.4, respectively.[27]

The SDD plotted in Figure 19b uses the same conditions as in Figure 19a, but also incorporates these additional proposed species. Furthermore and for comparison, the same two models were redrawn at experimental conditions suitable for ISE as well as polarography, *i.e.*, for $[M]_T = 1 \times 10^{-3}$ M and $[L]_T$:$[M]_T = 30$, as this would allow running a joint DCP/DPP/

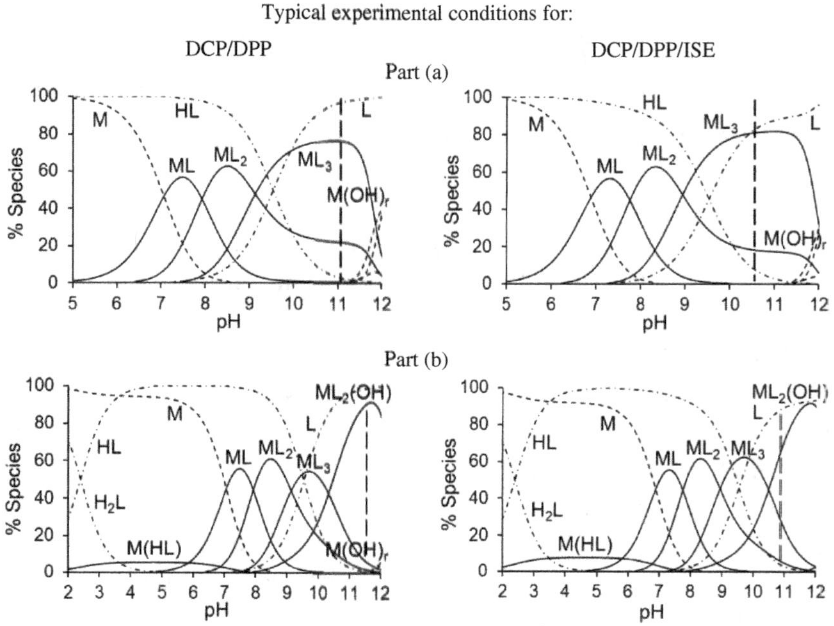

Figure 19. SDDs computed for the Cd(II)–glycine–OH (M–L–OH) system using experimental conditions suitable for DCP/DPP ($[M]_T = 1 \times 10^{-4}$ M and $[L]_T:[M]_T$ = 200) and combined voltammetric and potentiometric experiment, DCP/DPP/ISE ($[M]_T = 1 \times 10^{-3}$ M and $[L]_T:[M]_T$ = 30). The M–L–OH models include: (a) ML, ML_2, ML_3 and (b) additional M(HL) and $ML_2(OH)$ complexes. The vertical line shows the pH at which $Cd(OH)_2(s)$ precipitation is predicted. Unpublished data.

ISE experiment on a single solution sample; in this particular instance, highly comparable SDDs were obtained. Moreover, the predicted shifts in acquired $E_{1/2}$, E_p and E_{ISE} values are nearly indistinguishable; hence, a single Figure 20 is included as it predicts experimental output from all three techniques perfectly well and leads to identical conclusions.

Firstly, all complexes are predicted to form within the experimentally available window (prior to predicted precipitation) and, except for M(HL), they constitute large %-fractions of the total cadmium concentration meaning that their formation constants can be determined accurately by all techniques considered. From data shown in Figure 20 it follows that:

1. A very small shift of 1–2 mV (open circles) is predicted due to the presence of M(HL) and, unfortunately, it is hardly formed in both

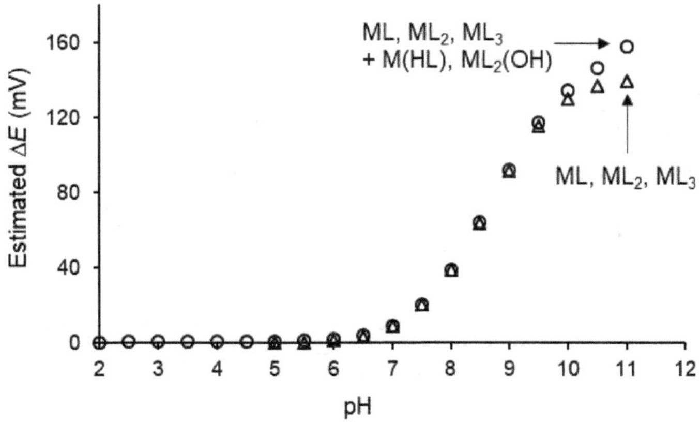

Figure 20. Estimated E-shifts for polarographic conditions when the ML, ML_2 and ML_3 model (triangles) and extended model with M(HL) and $ML_2(OH)$ (empty circles) was assumed. Unpublished data.

cases. Clearly, formation of this complex might be questioned under the experimental conditions used to draw the SDDs in Figure 18. In order to verify formation of M(HL), the best way forward would be to run more simulations in search for experimental conditions under which the M(HL) complex forms, if at all possible, with significantly larger %-fraction and somewhat larger E-shifts.

2. There is no doubt that it should be possible to detect the formation of $ML_2(OH)$ and determine its formation constant accurately. The difference between E-shifts at pH 11 reaches the value of ~18 mV and this is a significant value. This illustrates how formation of a species could be "missed" if the initial exploratory experimental conditions were too restrictive, *e.g.*, an experiment was terminated at pH ≤ 10.
3. If precipitation does not take place, as predicted in Figure 19, it would be highly recommended to collect more data points, at least up to pH of 12, to assure the highest quality refined formation constants. One must remember that precipitation might be a slow process and impossible to notice at its initial stage. The best way forward is to perform an out-of-cell and long-time-equilibration experiment to test for predicted precipitation. Moreover, when the intensity of a recorded polarographic signal is lower than expected after correction for dilution, then it might strongly suggest either the formation of an inert complex or precipitation.

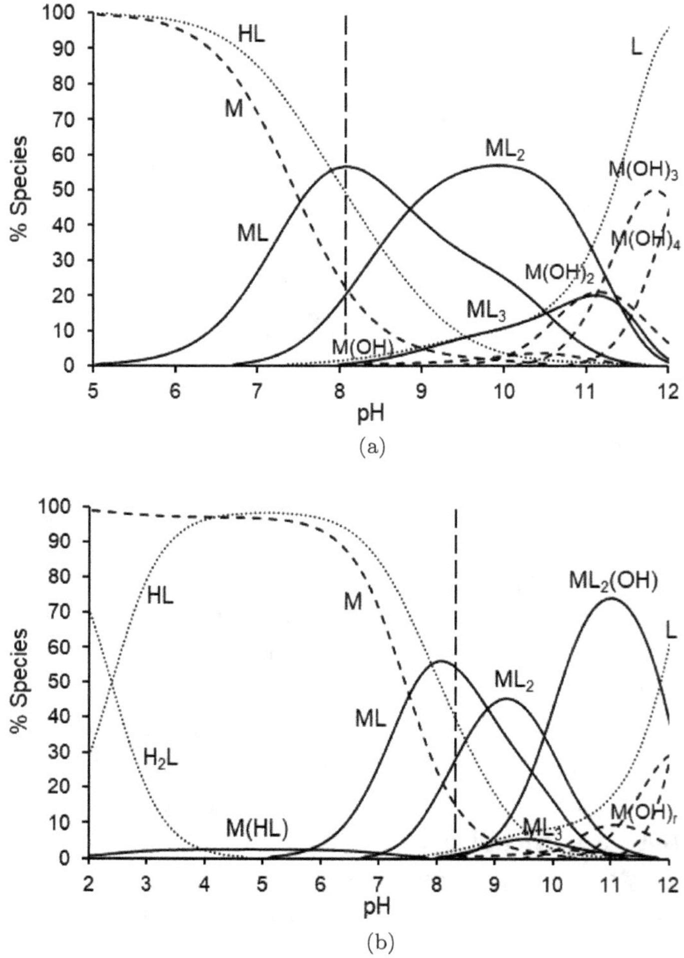

Figure 21. SDDs for the Cd(II)–glycine system using typical GEP conditions with $[M]_T = 5 \times 10^{-3}$ M and $[L]_T = 1 \times 10^{-2}$ M giving $[L]_T:[M]_T = 2$. The M–L–OH species models include (a) ML, ML_2, ML_3 and (b) additionally M(HL) and $ML_2(OH)$. Unpublished data.

When considering the SDDs for typical GEP conditions, where values of $[M]_T = 5 \times 10^{-3}$ M and $[L]_T:[M]_T = 2$ were used to plot Figures 21 for the two species models, a very different picture is seen. Precipitation is projected to occur at much lower pH in this case and only information regarding ML and possibly ML_2 would be obtainable under these conditions.

M(HL) is also barely formed under these conditions and even when experiments would start from pH 2, it would be very difficult, if at all possible, to detect this complex. This highlights the importance of using different techniques which probe different aspects of the M–L–OH systems and utilize different concentration ranges as well as $[L]_T:[M]_T$ ratios to generate complementary data. Although identical M–L–OH models may not be obtained from each experimental technique due to these disparities, the formation constants refined for the common complexes must be highly comparable.

It is also important to realize that the protocols described in this section do not consider any kinetics, be it homogeneous or heterogeneous. Thus, it is still possible for the data generated to be different to that expected from SDDs.

E. General Guidelines

A number of general concluding observations can be made in assisting the design of experimental conditions with the added reminder that there is always an interplay between all variables affecting a M–L–OH system. Having a large $[L]_T:[M]_T$ ratio suppresses the formation of polynuclear species. It also facilitates the complex formation reactions and shifts the formation of metal hydroxides to higher pH and hence postpones precipitation. At lower metal ion concentrations the extent of hydrolysis is also reduced and this should be taken into the account when studying highly acidic metal ions. Using smaller $[L]_T:[M]_T$ ratios means that certain species may go undetected because they do not form to a large extent in solution especially when using low metal ion concentrations, or because hydrolysis and precipitation occurred at lower pH compared to the case where higher metal ion concentrations were used. On the other hand, the amount of ligand added should not significantly contribute to the ionic strength of the solution and it is recommended to be at least ten times lower than the concentration of the supporting electrolyte.

Always of importance is the statistical verification of both the proposed model and the magnitudes of the stability constants. Thus doing pH titration experiments under different solution conditions is one approach, but the added use of different methodologies such as ligand titrations or

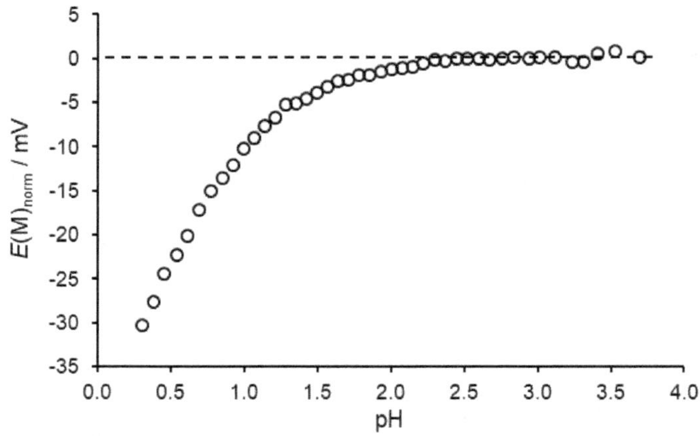

Figure 22. Normalized reduction potentials as measured in a cell containing a Ag/AgCl (3 M KCl) reference electrode in a salt bridge containing 0.5 M KNO_3 and submerged in a supporting electrolyte of 0.5 M HNO_3. The pH is increased by adding 0.5 M KOH. Unpublished data from Ref. 43.

employing other techniques such as ISEs or GEP adds impetus to the obtained results.

Finally, when working under very acidic conditions (generally below pH 2), one must be aware of significant pH-dependent variation of the diffusion junction potential (E_j), an unavoidable consequence of using a reference electrode irrespective of the technique employed. A potential difference arises when dissimilar solutions are in contact and is due to the difference in the rates of diffusion of ions across the boundaries where they are in contact, resulting in the two sides of the boundary being oppositely charged. This is particularly problematic when one of the solutions is highly acidic, since hydrogen ions have a very high mobility as compared to any other ions. Figure 22 shows the normalized reduction potentials determined for an uncomplexed metal ion (the identity of which is inconsequential) and any deviations from $E(\mathrm{M})_{\mathrm{norm}} = 0$ in the very acidic region is due to the diffusion junction potential, E_j.

Accounting for E_j is most problematic when pH titrations commence in the most acidic solution, where measured E(M) (in Eq. 17) would be offset by the value of E_j. Since E_j becomes smaller as the pH is increased, an error will be introduced when calculating $E(\mathrm{M}) - E(\mathrm{comp})_{\mathrm{pH}(i)}$. The magnitude of E_j will depend on the actual solutions in contact and can be estimated by the Henderson equation given Eq. 43.[51]

$$E_j = \frac{RT}{F}\frac{(U_1 - V_1) - (U_2 - V_2)}{(U_1' + V_1') - (U_2' + V_2')}\ln\frac{(U_1' + V_1')}{(U_2' + V_2')} \tag{43}$$

where $U = \Sigma c_+\lambda^\circ_+$, $V = \Sigma c_-\lambda^\circ_-$, $U' = \Sigma z_+c_+\lambda^\circ_+$, $V' = \Sigma|z_-|c_-\lambda^\circ_-$; and c, λ°, and z are the molar concentration, the limiting conductivity, and the magnitude of the charge for each of the cations and anions in solutions 1 and 2, respectively. A witness metal ion can be introduced in solutions used for polarographic pH titration measurements for studies below pH 2 to allow for the accurate compensation of E_j, and this will be briefly discussed in Section X.D. Note that diffusion junction potentials also limit the accuracy of the pH and ISE measurements in the very acidic region. In ligand titrations, where the pH is essentially kept constant, E_j will be negated when calculating $E(\mathrm{M}) - E(\mathrm{comp})_{x(i)}$ and would thus not impact on the results.

X. Practical Guide to Executing Experiments

Carefully planned and precisely executed experiments will provide the most reliable data and results. As such, some general experimental consideration will be highlighted before looking at more specific experimental procedures.

A. General Considerations for Experiments

Since stoichiometric equilibrium constants are of interest in this work (*i.e.* formation constants calculated using the concentrations of the species), all experiments must be performed at a fixed ionic strength and temperature. Maintaining a fixed ionic strength requires the addition of a large concentration of inert salt such that any changes in concentrations of the reacting species do not significantly affect the overall ionic strength. Ideally, the concentration of the inert electrolyte should be 100 times greater than that of the reacting species. This may, however, not always be achievable when wanting to work under particular conditions, for example under physiological conditions where the ionic strength is 0.15 M (H,Na)Cl. When using polarography, the metal ion concentration can be very low (~10^{-4} M or even less). Even though the ligand concentration is generally ~10^{-2} M, the concentration of the ligand that is not involved in complex formation reactions does not change significantly and this is due to the low metal ion concentration and large excess of the ligand. This means that the ionic strength will vary mainly due to the extent of deprotonation of the ligand. It is of paramount importance to understand that it is almost an impossible task to keep an ionic strength at an exact value throughout an experiment. However, keeping ionic strength as close as possible to the assumed (required) value is in accord with a good analytical practice. As constancy of ionic strength is an ambiguous requirement, the best practice would be to report the range of ionic strength within which an experiment was performed. Hence, keeping and reporting detailed experimental conditions should be seen as not negotiable requirement. The inert electrolyte has the added benefit of being the supporting electrolyte necessary for polarographic measurements. It reduces the ohmic drop and ensures a diffusive mass transport mechanism for the studied species. The temperature of the solution is normally maintained by the use of a jacketed cell connected to a thermostatic water bath and the temperature is best monitored using a thermocouple.

All solutions should be made using deionized water (18 MΩ cm) and high purity reagents. All cell components must be clean and dry so that no dilution of the solutions occurs, thus concentrations are "exactly" known. For titration experiments it is recommended that an anti-diffusion tip be used to deliver the titrant to avoid leakage into the test solution with time.

All ion sensing electrodes (be it a GE (pH electrode) or metal ISE) should be in good condition and stored according to the manufacturer specifications to ensure fast and stable responses. Note that to prolong the lifetime of the GE, it should not be left in alkaline solutions for extended periods as this will corrode the membrane. The use of a combined GE-reference electrode is suggested to reduce the number of probes inserted into the cell when running full polarographic experiments. Since solution pH is a critical factor in obtaining information to solve the required mass balance equations, the pH of all solutions is measured, irrespective of the technique used. Other ISEs generally require a separate reference electrode.

Most ISEs for metal ions have a solid crystal membrane. These electrodes are prepared by gently polishing the surface with Al_2O_3 paste, then rinsing and soaking in 0.1 M metal ion solution for at least 1 hour until a constant reading in its potential is attained. Before use and between measurements, the electrode is stored in a metal ion solution with a total concentration similar to that of the sample solution. In some cases, it may be beneficial to include the ligand being studied in the solution just before use for added stability (preferably at a pH where minimal complexation occurs thus having both the free metal ion and the ligand in solution); this must be tested first.

Manufacturers normally provide specifications for ion sensing electrodes, giving the pH range in which the electrode is stable and the linear calibration range, but it is still important to regularly establish the electrode performance. For metal ISEs in particular, the stability across the pH range of study must be checked by ensuring the response of ISE (E_{ISE}) remains constant (within experimental error) for metal ion solutions of the same concentration but at different pH. Limitations in the acidic region are due to the diffusion junction potential of a reference electrode and the dissolution of the crystal membrane, and the basic region is limited by the hydrolysis of the metal ion. The long-term stability of the ISE also needs to be tested as experiments usually take several hours. This is achieved by periodically taking a potential reading with the ISE in a solution containing (i) just the metal ion, and (ii) the metal ion with the ligand using concentration ranges that would be typical for the experiment to be performed. This can also be done at a number of pHs and should once again provide stable values. This is particularly important when very high ligand concentrations are used which may lead to adsorption on the electrode surface resulting in slower response times and altered electrode response characteristics.

All ion sensing electrodes must be meticulously calibrated using the suggested procedures which are described in following sections. The response slopes should always be close to the theoretical Nernstian value for good performing electrodes. It is recommended that calibrations are done before and after each titration experiment (which run for many hours) to assure the integrity of the electrode is maintained throughout. To account for slight drift, the average of the two calibrations is used.

B. Glass Electrode Calibration

The GE calibration must be performed under conditions that reflect the polarographic experiment as closely as possible for the calibration to be most applicable. Employing buffer solutions to calibrate the electrode does not provide sufficient accuracy, as the ionic strength of these solutions are often not controlled, a two-point calibration is not accurate enough, and the hydronium ion activity rather than its concentration is determined.

The recommended procedure for the calibration of a GE involves the titration of a strong acid by a strong base.[52,53] These are chosen such that the counter ions are the same as those for the supporting electrolyte. For example, if a KNO_3 supporting electrolyte is used, HNO_3 and KOH would be employed. The ionic strengths of strong acid and strong base solutions should be adjusted using the supporting electrolyte (which should be the same for both solutions) and these solutions must be standardized. The basic solutions should be standardized at least every two weeks as these solutions could react with CO_2 from the atmosphere.

The base solution is then incrementally added to a fixed volume of the acid solution while continuously stirring, ensuring that sufficient data points are obtained across the pH range. For the case where the acid and base solutions have the same concentration, the final volume of the base required is double that of the initial volume of the acid if an equal number of data points in the acid and base regions is desired. The solutions are also continually purged with high purity inert gas (either nitrogen or argon) to exclude CO_2 from the cell, as well as to mirror the conditions used for the polarographic experiment as will be discussed. At each step in the titration, the volume of base added and the stabilized potential reading of the GE is noted. When using an automated system, as described in Section XI, the stability of the GE potential is tested by applying a dedicated algorithm. For example, a short waiting period (~15 s) is allowed after the titrant addition, then the GE potential is acquired every 2 s. After 10 readings, the stability criterion is tested and only one out of ten values is allowed to deviate by a maximum of 0.1 mV. If this is not met, another measurement is made and the stability criterion is retested by considering the last ten acquired data points. A dedicated file is generated that contains all readings after each titrant addition. This is very useful information as slowly responding and unstable electrodes can be identified and replaced before preparing a solution sample for the intended M–L equilibrium study.

Knowing the volumes and concentrations of the acid and base at each titration step, the pH can then be determined (calculated using –log $[H^+]$) and the GE potential (E_{GE}) *vs.* pH calibration plot drawn. It should be noted that calculations for the basic region require the K_w value which should be obtained for the working temperature and ionic strength (and for the specific supporting electrolyte if available). The pK_w value of 14 that is widely used is only applicable at 25 °C for pure water (*i.e.* 0 M ionic strength) and is more accurately 13.997 ± 0.003.[12,54] At 0.1 M ionic strength (more realistic experimental conditions) pK_w is 13.95 and 13.78 at 20 and 25 °C, respectively.[12,54] At significantly higher ionic strength (~8 M), the pK_w value would be about 2 log units higher.

An example of a calibration plot is given in Figure 23. Deviations from linearity in the low buffer region are generally observed. This could be due to small errors in standardized solution concentrations,[55] the presence of impurities,[52] or having a glass vessel behaving as an ion-exchanger.[56]

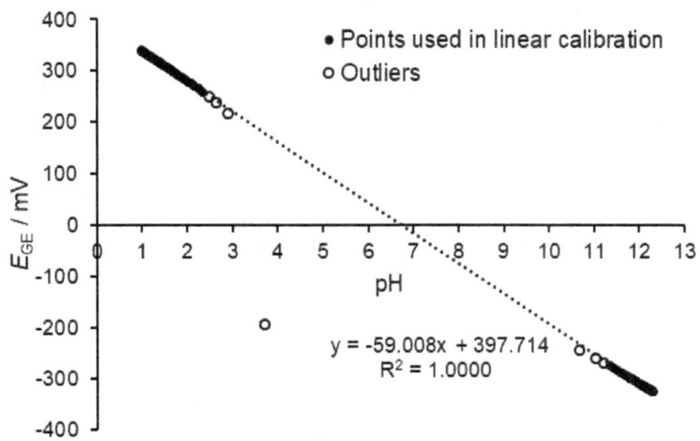

Figure 23. Glass electrode calibration graph as determined from the titration of 0.1008 M HNO_3 by 0.1003 M NaOH, both adjusted to 0.50 M ionic strength using $NaNO_3$, at 25.0 °C. An initial volume of 25.00 mL HNO_3 was used and NaOH was added in 0.50 mL increments up to a final volume of 50.00 mL. Unpublished data.

Regardless of the cause, it is good practice to omit the few data points which clearly deviate from linearity and the best straight line is then fitted through the remaining data from both the acidic and basic regions. The straight line represents $E_{GE} = slope \times \text{pH} + E°$, where the *slope* should be close to the Nernstian value at specific temperature. The E_{GE} value at pH 7 should also be close to zero and lie within the manufacturer specified limits.

For experiments where solutions also include the metal ion and ligand of interest, the GE response is directly measured and using the calibration data, the pH and hence $[H^+]$ of the solution can be determined.

C. Polarographic Experiments

Mercury electrodes are ideal when working in acidic solutions due to the high hydrogen overpotential, but the anodic region is limited by mercury oxidation. The benefit of using a dropping mercury electrode (DME) is that the surface is uniform, reproducible and is continually renewed so that oxidation or reduction products do not build up and change the character of the electrode surface. Notably, no solid or film electrode offers the same long-term reproducibility required in the multi-hour titration experiments employed. Unfortunately, with mercury being a toxic and hazardous pollutant, more strict control of its use has been implemented making it more difficult to obtain and the correct precautions need to be taken during its use and disposal.

The parameters used in the polarographic titration experiments should be chosen in such a way as to provide a balance between the various requirements. For example, sufficient time should be given for the solution species to reach equilibrium at each point in the titration and a sufficient number of data points must be obtained to allow for accurate analysis, yet also keeping the total time of the experiment as short as possible to ensure the best GE performance and minimal evaporation caused by prolonged purging. The polarographic techniques, such as DCP (it is implemented as sampled-DCP in PC-controlled instruments) and DPP, can be used where DPP is more sensitive and recommended for systems that are electrochemically reversible. However, for systems where the electron transfer rate is sluggish, resulting in quasi-reversible responses, DCP is recommended as I_d is independent of the degree of reversibility and the reversible half-wave potential can be extracted from a quasi-reversible wave (see Section XII.C)

The parameters selected for the polarographic measurements should ensure minimal noise in the data, especially when low concentrations of the metal ion are used. Typical measurement parameters for DCP are: a step potential of 4 mV, a drop time of 2 s and a current integration time of 60 – 100 ms. In the case of DPP, typical parameters include: a drop time of 2 s, pulse and step heights of 50 and 4 mV, respectively, and pulse width and integration time of 200 and 90 ms, respectively. Again these parameters need to be carefully assessed as, for example, a 50 mV pulse height improves the sensitivity of the measurement, but can only be employed for M–L systems where fast homogeneous equilibrium can be attained. Preferably, the potential window in which the measurements are made should allow for "clear" baselines on each side of the DPP peak or DCP wave. This means that one should avoid, if at all possible, working in the region of applied potentials where mercury electrode dissolution or hydrogen evolution occurs. This will not only allow for more accurate determination of $E_p/E_{1/2}$ and I_p/I_d values, but it will also preserve solution equilibria from external disturbances. Typically, this window needs adjusting as the titration experiment proceeds and particularly so for cases where large shifts in potential occur due to complex formation.

A three electrode polarographic cell is used containing a DME as the working electrode, a counter electrode (such as platinum) and a reference electrode (such as Ag/AgCl/3 M KCl). For systems not studied in chloride solutions, it is recommended that a salt bridge (containing the supporting electrolyte used) be introduced between the reference electrode and the test solution in order to prevent chloride contamination. The cell should additionally contain the combined GE, the thermocouple, inlet for the titrant and the purge gas.

Since oxygen itself is electroactive, all solutions need to be thoroughly purged using a high purity inert gas and depending on the gas flow rate, this takes about 1 minute per mL of solution. The gas is normally humidified (by bubbling it through water when working in aqueous solutions) before it reaches the cell to minimize evaporation of the sample solution as far as possible. This time also allows for the solution to reach thermal and chemical equilibrium. It is important that there is a negligible change in the total metal ion concentration due to its reduction at the DME during a polarographic measurement; hence, the sample solution volumes should not be too small and thus 20 – 25 mL is generally the starting volume.

The procedures given below are for accurate experiments, so the conditions can be slightly relaxed when performing the first exploratory experiment. This is usually a pH titration across a wide pH range in order to verify any predictions made so fewer polarograms can be collected. Measurements are normally not made beyond pH ~12 to maintain the GE performance and the experiment is stopped once precipitation is observed. The polarograms must be analyzed to check for reversibility to ensure the correct technique and parameters are used for the measurements; this is discussed in depth in Section XII. The data can then be analyzed to produce a more reliable species model and log β values for further investigation using SDDs to optimize the experimental conditions for a more "perfect" experiment.

Both pH and ligand polarographic titration experiments have a number of common steps in their procedures, namely:

1. The experiment is started by collecting a polarogram of the deoxygenated background solution to ensure that no impurities (this includes dissolved O_2(g)) are present in the solution and to assure that an adequate purging time is being utilized. A current maxima suppressant has to be added and a few grains of gelatin is sufficient. The background solution is either the same as the acid solution used for the GE calibration or a mixture of the acid and base solutions to give the required pH, with the supporting electrolyte to fix the ionic strength.
2. Solutions should always be stirred and purged during and after any addition of a titrant to ensure proper mixing, sufficient deoxygenation and equilibration before polarographic measurements. It is critical that prior to recording a polarogram the species composition at the working electrode surface should represent that in the bulk solution as closely as possible. Thus stirring and purging are only stopped at least 20 s prior to and during acquiring the polarographic signals to assure that the only mode of transport is diffusion.

3. GE potential readings are always taken just before a polarogram is recorded; this is to assure consistency with the methodology used in the GE calibration process.
4. Temperature readings should be taken at the same time to monitor the variation (if any) throughout the experiment.
5. A stock solution of the metal ion of interest is required, having a known concentration of ~0.1 M and stabilized by 0.5 – 1 M acid.
6. Small volume (of the order of microliters, typically between 5 – 200 μL) of the metal ion stock solution is added to the background electrolyte solution to obtain the required concentration of the metal ion.
7. A minimum of three polarograms for the reduction of the free metal ion should be collected (with a period of stirring and purging between each measurement) to check that there is no variability due to instrument issues or insufficient initial purging. Since the shift in potential is always calculated relative to the $E_{\mathrm{p}}^{\mathrm{r}}(\mathrm{M})$ (or $E_{1/2}^{\mathrm{r}}(\mathrm{M})$) value (Eq. 17) it is critical that this value be accurately determined.

1. *pH titrations*

The ligand is added to the metal ion solution either as the solid or as a small volume of a 0.1 – 1 M stock solution. This should give the sought-after $[\mathrm{L}]_{\mathrm{T}}$: $[\mathrm{M}]_{\mathrm{T}}$ ratio and a polarogram is recorded after a stable pH reading is established following introduction of the ligand to the sample solution.

The pH titration is then commenced by adding the base stock solution (as was used for the GE calibration) until a change in pH of ~0.05 – 0.1 is obtained. This gives 10 – 20 data points (polarograms) per pH unit that will facilitate modelling of M–L–OH system and refinement of the formation constants. The magnitude of the incremental pH change depends on the size of the pH range being studied (as decided from analysis of the SDDs) and the smaller the pH range, the smaller the incremental pH change should be. The aim is to acquire about 40 – 60 polarograms in total to produce good overall modelling statistics.

When the sought after change in pH is reached, additional time is allowed so that the measurements made truly represent the metal–ligand system at equilibrium. A similar stability criterion, as applied for the GE calibration, can be used as an indicator of equilibration in solution. Once the pH is deemed as stable, a polarogram is recorded. This procedure continues until the final pH value is attained. Understandably, this protocol is not applicable for M–L systems that undergo slow (above 20 minutes) equilibration and out-of-cell experiments should rather be considered. Slow equilibration or essentially no complex formation on the time scale of the experiment can be identified by unchanging polarographic signals (other than from dilution effects) across a wide pH range.

2. *Ligand titrations*

In this case the titrant is a standardized stock solution of the ligand (~0.1 – 0.5 M; the higher the concentration the better, as discussed in Section IX.C). The pH of the titrant should be adjusted (using the strong acid or base solution) to the same value as the pH of the initial background electrolyte solution in the cell. The ligand solution is then accurately added to the test solution to give an initial $[L]_T$:$[M]_T$ ratio (generally ~10 – 20) and after equilibration, all measurements are made. Additions are repeated until the final $[L]_T$:$[M]_T$ ratio is reached, as decided from the information gained from the SDDs. The increment of ligand solution added must be such that there is a minimum of 30 polarograms collected. One must be aware that there might be a problem due to low solubility of fully protonated ligands. It is then a good practice to check whether assumed $[L]_T$ can be reached at the selected pH before titration.

D. Experiments Under Very Acidic Conditions

Some pointers will be made in this section rather than providing a complete methodology on how to study complex formation under very acidic conditions, where studies down to pH 0.3 are considered. One among biggest challenges is the diffusion junction potential (E_j) discussed in Section IX.E, as it is present wherever a reference electrode is employed. This becomes a real challenge when experiments are performed by varying the pH since E_j also varies with pH, especially below pH 2 and above pH 12 with the actual magnitudes of E_j depending on the compositions of the solutions in contact. This has implication for both the GE calibration and the polarographic pH titration experiments.

Considering a combined GE, the reference filling solution (3 M KCl) is in direct contact with the test solution. For GE calibrations starting at pH 0.3 (a 0.5 M H^+ solution), the linear pH-range is generally between about pH 0.6 and 12.8, but this varies slightly due to the glass membrane condition.[42] To produce the most accurate relationship between the pH and GE potential, a quadratic calibration is used in the very acidic region to account for the curvature until this function is intercepted by the straight line calibration which is then used for the remaining pH range.[42] Curvature is also noticed at the highest pH values due to the alkaline error and E_j. This region is rather avoided, however, to prevent corrosion of the glass membrane.

For the polarographic pH titration experiments, a salt bridge is often used between the reference electrode and test solution, and it is the E_j between the test and salt bridge solutions that varies as the test solution is titrated. The value of E_j can be estimated (Eq. 43), but the most accurate approach is to incorporate a witness metal ion in the test solution itself. This witness ion should not undergo complex formation with the ligand being studied, especially in the very acidic range, such that any change in the reduction potential of the witness metal ion can be ascribed to the change in E_j alone. Thallium(I) has been successfully used as a witness ion, but its high toxicity should be noted and should be treated accordingly.[43,44]

Lastly, further considerations are required when studying complex formation of very acidic metal ions. For example, Bi(III) undergoes hydrolysis below pH 1 already and precipitates at ~pH 2 as a Bi(III)-oxy-anion species (where the anion is from the background electrolyte). This means that even at very low pH $E_p^r(M)$ (or $E_{1/2}^r(M)$) cannot be directly measured as hydrolised Bi^{III} species have already formed. The shifts in potential due to the formation of these species can, however, be calculated using the formation constants of the hydroxo species (Eq. 42) and accounted for. In the case of Bi(III), formation of nitrate species (from the background electrolyte) also had to be accounted for in the model, making it a complicated system to study, yet achievable when using polarography with the experimental and modelling tools available.[45,46]

E. Ion Selective Electrode Calibration

A metal–ligand equilibrium (MLE) experiment involving ISE can be executed as a separate potentiometric titration or, provided that analytical conditions are conducive for the use of ISE, as a combined (a) ISE-polarographic, (b) ISE-GEP, or even (c) combined GE-ISE-polarographic experiment. Regardless, in each case the ISE must be calibrated to verify its performance and suitability for the planned MLE experiment.

Calibration of ISE is very different from that described for a GE and it can be conducted in two ways. A series of solutions with exactly known concentration of the free metal ion, typically in the concentration range between 0.1 and 1×10^{-4} M, must be freshly prepared. All calibration solutions must be adjusted to a specific pH and required ionic strength using a background electrolyte. The ISE should be rinsed with water and gently dried before each measurement. Preferably, six (or more) measurements should be taken, starting from least concentrated solution. Having completed calibration measurements, electrode should be placed in a solution with the free metal ion concentration comparable to the initial $[M]_T$ envisaged for the sample solution; the ISE is ready for the M–L–OH equilibria studies when its potential does not drift in time. An example of a four-point calibration performed before and after a M–L–OH equilibria study experiment is shown in Figure 24a. This can be rated as good performing Cd^{2+} ISE as (i) its response slope is close to the Nernstian value

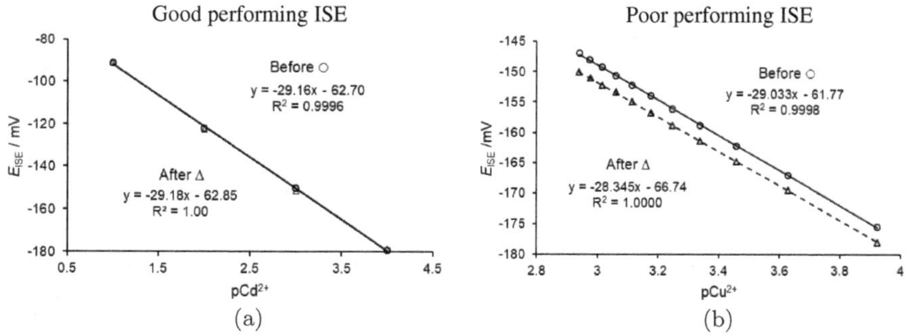

Figure 24. Examples of ISE calibrations in 0.5 M $NaNO_3$ at 25 °C, before and after a titration experiment. Part (a) shows a good performing ISE calibrated in the range ~0.1 to 1.0×10^{-4} M Cd^{2+}. Part (b) shows a poor performing ISE calibrated in the range ~1.0×10^{-3} to 1.0×10^{-4} M Cu^{2+}. Adapted from Ref. 58.

and hardly changed (increased by 0.02 mV) and (ii) the E° value (which represents an expected potential in 1 M metal ion solution under the experimental conditions used) has increased just by 0.15 mV. In such a case one should use all the eight points for computing the linear regression equation.

Data shown in Figure 24b illustrates another calibration protocol where titration of a background electrolyte with a concentrated stock metal solution was performed. Naturally, the titrant and titrated solutions must be adjusted to the same pH and ionic strength. Titration data (circles) collected before the experiment involving M–L–OH equilibria shows a very good performance of the Cu-ISE. Unfortunately, there is clearly some drift in the electrode response after the experiment, as indicated by the variation in both the response slope and intercept of the calibration. Driven by curiosity, one might consider using regression data from both titrations in order to find the impact it makes on the M–L–OH model and refined formation constants. There might be many reasons for such a difference in the calibration data. Besides the potential impact of the MLE experiment on the ISE surface, a reference electrode and how it was used for both calibration titrations and between should be carefully considered. This clearly highlights the importance of reassessing the calibration after the titration experiment.

XI. Recommended Instrumentation, Automation and Data Collection for Potentiometry and Voltammetry

Speciation methods discussed herein involve GE- and/or ISE-based potentiometric measurements, polarographic measurements (DCP, DPP, or other voltammetric modes), and any feasible combination of potentiometric and voltammetric measurements performed on a sample solution. In polarographic studies of M–L–OH systems, either a sample solution containing a specific concentration of a metal ion is titrated with a ligand solution at fixed pH or a solution of fixed $[L]_T$:$[M]_T$ ratio is titrated with a base to vary pH. Polarograms are recorded at various pH values or ligand concentrations. Typically, the variation in solution composition has been achieved by manual titrant additions and polarograms recorded with the aid of a computer. In this regard, these polarographic experiments have mainly been computerized rather than automated. Inherently, these titrations can be highly tedious, time-consuming, susceptible to many, and difficult to trace human errors.

For rigorous analysis and refinement of M–L–OH models, as many experimental points as possible should be collected in the experimentally accessible pH range. In this regard, a computer-controlled instrumental setup, capable of controlling the execution of the titration and acquisition of data from potentiometric and polarographic probes, would be advantageous. The benefits of automation include (i) improvements in accuracy, precision and reproducibility, (ii) elimination of time-consuming and tedious manual processes, and (iii) minimization of potential human errors. Furthermore, information obtained by using automated procedures in adjustment of solution composition and data capture of potentiometric or polarographic measurements is usually not conveniently obtained at the desired frequency by manual procedures. Finally, digitally stored experimental data is perfectly suited for evaluation, validation, processing, *etc.*, and can be shared, exchanged as well as conveniently archived for years.

Research-dedicated electrochemical workstations equipped with programmable automated procedures and automated titration apparatus can either be sourced from specialized manufacturers or specifically developed for dedicated research purposes.[40,58] Since most hardware components required for potentiometry and voltammetry can be sourced separately from different manufacturers, a virtual instrumentation (or modularized instrumentation) approach can be adopted to create user-defined instrumental solutions for solution chemistry studies. In this case, stand-alone hardware components (digital burettes, potentiostats, valves, *etc.*) perform different tasks imposed on them by dedicated computer software modules.

A block diagram of a typical instrumental setup designed for automated potentiometric and polarographic measurements is presented in Figure 25a. The basic hardware components are:

1. A personal computer (PC) equipped with appropriate control software modules to control entire instrumentation, execute all functionalities, digitally generate DC/DP or any required voltage ramps, acquire and store all experimental data.

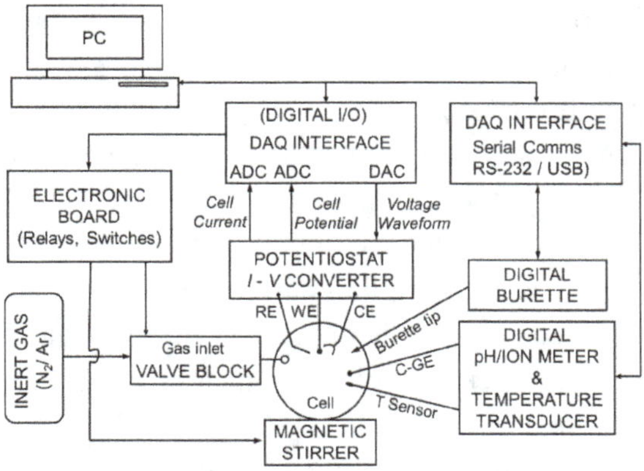

(a)

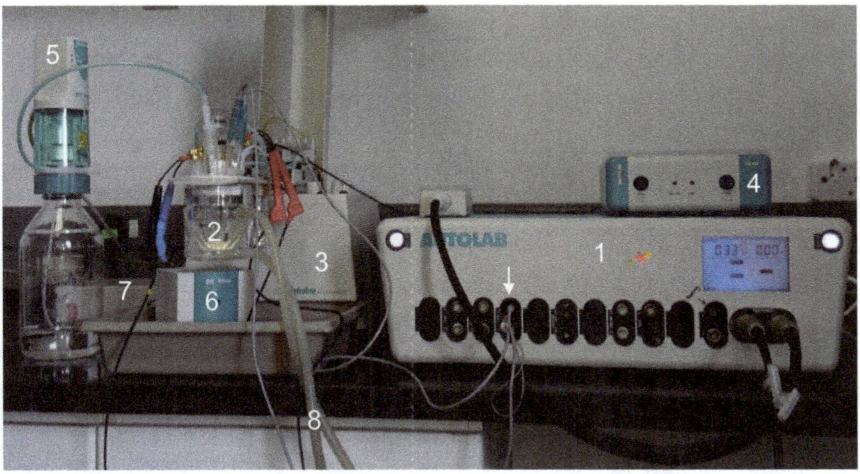

(b)

Figure 25. (a) A block diagram of an instrumental setup for combined potentiometric and voltammetric measurements on a single sample solution. PC: Personal Computer; DAQ: Data acquisition; ADC: Analog-to-Digital converter; DAC: Digital-to-Analog converter; I–V: Current–voltage; Serial Comms stands for Serial Communication; WE: Working electrode; RE: Reference electrode; CE: Counter electrode; C-GE: Combination Glass Electrode; C-ISE: Combined Ion-Selective Electrode; I/O: Input/Output; T Sensor: Temperature sensor (or thermocouple). (b) A photograph of an electrochemical workstation used for automated titration procedures which incorporates: 1 – potentiostat with built in pH meter (indicated by the arrow), 2 – electrochemical cell, 3 – cell stand, 4 – interface to the cell stand (to control mercury electrode and gas flow), 5 – dosing device; 6 – magnetic stirrer, 7 – interface to the dosing device and stirrer, 8 – tubing to a thermostat-controlled water-bath.

2. Data acquisition (DAQ) interface is used for voltammetric measurements. It is equipped with:
   (a) Analog-to-digital converter (ADC) channels with at least 16-bit resolution for data sampling at a fraction of a mV. One ADC channel is dedicated to sampling the applied to the working electrode potential (as controlled by the potentiostat) and a separate ADC channel is used for sampling current responses to the applied potential (current flowing between the working and auxiliary electrodes) *via* a current-to-voltage converter;
   (b) Digital-to-analog converter (DAC) channel (with at least 16-bit resolution) for applying a voltage waveform to the electrochemical cell *via* the potentiostat.
   (c) Digital input/output (I/O) channels. They are used for digitally controlling switches or relays linked to "on-off" operations of gas valves on the valve block of the voltammetric stand and the magnetic stirrer.
3. A potentiostat with current-measuring and/or signal conditioning circuits (current–voltage converter and/or integration amplifier).
4. A voltammetric stand consisting of a valve block (with digitally controllable solenoid valves) used for control of gas flow into the electrochemical cell and control of dropping mercury electrode functionality in polarographic measurements.
5. An electronic board with digitally actuated relays or electronic switches.
6. Magnetic stirrer (digitally controllable).
7. Digital pH/ion meters for potentiometric measurements (controlled *via* a dedicated DAQ interface using serial communication protocols involving RS-232 or Universal Serial Bus, USB), which are used to sample actual readings displayed on the pH/ion meters.
8. Digital burette for volumetric titrations (controlled *via* a DAQ interface using serial communication protocols).
9. A temperature sensor or thermocouple (typically incorporated in a multipurpose digital pH/ion meter equipped with a temperature transducer).

The main features of an automated setup designed for combined voltammetric and GEP and/or ISE titrimetric measurements include:

1. Flexibility to run independent GEP/ISE potentiometric titrations with automated data capture of added titrant volume and measured GE/ISE potential (mV) readings at each titration stage. This is a useful feature invoked prior to M–L experiments; *e.g.*, (a) in calibration of the GE by a standard acid–base titration, and (b) in acid (or base) titration of a ligand to establish its protonation constants.
2. Flexibility to invoke voltammetric techniques (DCP, DPP, *etc.*) independently, *e.g.*, to record the free metal ion polarograms before the titration is commenced or to study the behavior of a particular ligand in a given solution.
3. Titrant addition by user-defined fixed volume increment or a dynamically determined volume increment.
4. After each titrant addition the exact volume of titrant is captured and the change in solution potential (as a function of time) is followed potentiometrically (with the GE as the main probe to detect solution equilibration).
5. GE and/or ISE equilibrium potentials are programmatically monitored by special algorithms for detecting constancy of the mV readings from the digital pH/ion meter (constrained to a certain user-defined criterion, such as standard deviation of the average of successive readings). The established potential (an averaged value meeting the criterion for equilibration as a function of sampling time) is taken as the final equilibrium potential (mV) at a particular titration stage.
6. Polarograms are not necessarily recorded after each titrant addition, but rather the frequency is programmatically determined by a user-defined rate based on overall pH changes.
7. Flexibility to record polarograms within a user-defined fixed potential window, or dynamic adjustment of the voltammetric potential window, based on special algorithms, which detect any significant fluctuations in recorded voltammograms as the solution composition changes.
8. Online detection of shifts in $E_{1/2}$ in recorded DC waves or E_p of DP curves as pH is increased during successive DCP/DPP measurements at variable pH.
9. Implementation of multiple end-of-experiment conditions using either (a) a combination of user-defined parameters, such as, pH meter and

burette readings at any stage of the experiment (measured GE potential, ISE potential, measured pH or titrant volume) or (b) operator's intervention *via* an overriding stop command.

XII. A Practical Guide in Evaluation of Polarographic Signals

Departure from electrochemical reversibility is a common phenomenon and can be observed for a metal ion as well as metal complexes. Furthermore, electrochemical reversibility can vary with experimental conditions and solution composition when pH or ligand concentration is varied. Therefore, for polarography to be a diverse and useful technique in speciation studies, it is important to have methodologies that can interpret M–L–OH systems involving not only electrochemically reversible reactions, but also those systems that exhibit deviations from reversibility.[40,41,59]

As already stressed, the refinement operations that are invoked to determine formation constants require use of data coming from reversible processes. Hence, it is imperative to test for electrochemical reversibility prior to any modelling of M–L–OH systems and refinement operations employed to compute formation constants. Suitable reversibility tests and extraction of relevant DPP E_p^r and I_p^r as well as DCP $E_{1/2}^r$ and I_d parameters from experimental curves are described in detail below. Also and importantly, protocols useful in computing reversible half-wave potentials from quasi-reversible polarograms will be discussed.

A. Polarographic Signals

The experimentally recorded polarographic current I^{exp} at any applied step potential E_{step} is made of two major components, namely, reduction current I^{red} and background current I_b and, as an example for a DPP current, we can write:

$$I_{DPP}^{exp} = I_{DPP}^{red} + I_b \tag{44}$$

Components of the total experimental DPP and DCP currents are presented in Figure 26. The main goal in curve-fitting (non-linear least

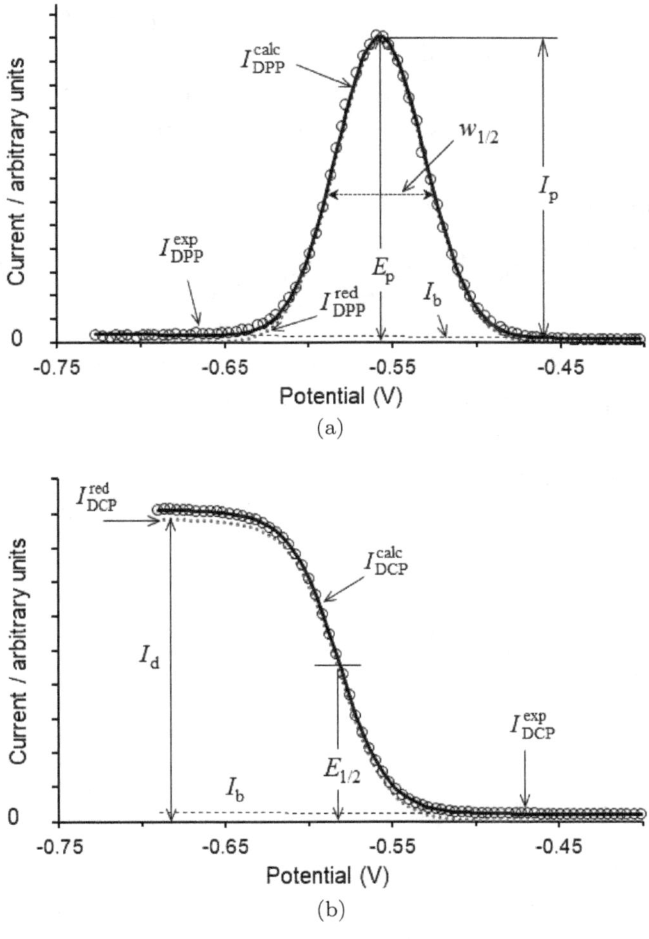

Figure 26. Schematic presentation of components extracted from fitting experimental polarograms: open circles represent a total current at each applied step potential, the solid line is the calculated total current, the dotted and dashed lines stand for the reduction and background current, respectively. Part (a) – a DPP curve; part (b) – a DCP wave. Unpublished data.

squares fitting or regression analysis employing a mathematical model function) is to closely match the calculated total currents (I^{calc}_{DPP} or I^{calc}_{DCP}) to the experimental currents (I^{exp}_{DPP} or I^{exp}_{DCP}) at every potential at which the current was sampled. To this end, not only is separation of the faradaic reduction and background signals achieved, but also the associated polarographic parameters (I_p, E_p and $w_{1/2}$ of a DPP peak, or I_d and $E_{1/2}$ of DCP wave) are also obtained.

Using initial best guesses of polarographic parameters in the model functions, these parameters are refined using applicable least squares algorithims[60] by minimizing the objective function, *i.e.*, the sum of squares of errors, as given in Eq. 45:

$$\sum_{1}^{n}\left[I^{\exp}\left(E_{\text{step}}\right)-I^{\text{calc}}\left(E_{\text{step}}\right)\right]^{2} \tag{45}$$

where, for each nth applied step potential E_{step}, a calculated total current I^{calc} is compared against the experimentally recorded current I^{exp}. Typically, I_b can be approximated by either linear, polynomial, or exponential functions, Eqs. 46–48:

$$I_b = a + bE_{\text{step}} \tag{46}$$

$$I_b = a + bE_{\text{step}} + c(E_{\text{step}})^2 \tag{47}$$

$$I_b = a + bE_{\text{step}} + c[\exp(dE_{\text{step}})] \tag{48}$$

where a, b, c, and d are fitting parameters in a given model function of I_b.

B. Analysis of DP Polarograms

An advantage of DPP is that specific peaks can be observed for individual electrochemical processes with peak potentials differing by as little as 0.05 V, in contrast to classical DC techniques, which require a potential breadth of about 0.2 V for good wave resolution. DPP has a better ability to discriminate against capacitive current, because it measures a differential current (helping to subtract any residual capacitive current that remains prior to each pulse step). Limits of detection with DPP are typically in the range $10^{-8} - 10^{-9}$ M.[61] In DPP studies of M–L–OH systems, the homogenous and electrochemical pathways involved (Scheme 1) may result in well-resolved characteristic reversible peaks, broad non-reversible peaks, overlapped reversible peaks or overlapped non-reversible peaks. The strategy for extraction of relevant peak parameters could require rigorous fitting of polarograms, using complicated and involved models constructed from voltammetric theories. These rigorous models account for main electrochemical mechanisms of appropriate electrode reactions, diffusional flux of the electrochemically active species and relevant thermodynamic and kinetic parameters.[59]

A useful and much simpler strategy involves building a simple curve-fitting model to reproduce a DPP reduction current $I_{\mathrm{DPP}}^{\mathrm{red}}$. The modified Gaussian function (Eq. 49) can be used to deconvolute the reduction current as a function of the applied E_{step} of typical "bell-shaped" DPP peaks.[24]

$$I_{\mathrm{DPP}}^{\mathrm{red}} = I_{\mathrm{p}} \exp\left(\frac{-\left(E_{\mathrm{step}} - E_{\mathrm{p}}\right)^2}{2\left(\frac{w_{1/2}}{2.345}\right)^2} \right) \tag{49}$$

This "$y = f(x)$" function, where $y = I_{\mathrm{DPP}}^{\mathrm{red}}$ and $x = E_{\mathrm{step}}$, contains two parameters needed in computing $CS_{\mathrm{pH}(i)}$; namely, E_{p} and I_{p} as well as the half-peak width $w_{1/2}$ parameter that can be used to monitor electrochemical reversibility.

A curve-fitting model (Eq. 50) for calculating E_{p}, I_{p}, and $w_{1/2}$ from an experimental DP polarogram incorporates Eq. 49 for the calculated DPP reduction current and an applicable I_{b} function:

$$I_{\mathrm{DPP}}^{\mathrm{calc}} = I_{\mathrm{DPP}}^{\mathrm{red}} + I_{\mathrm{b}} \tag{50}$$

Evaluation of electrochemical reversibility, as reflected in a given DP polarogram, can be done by inspection of the magnitude of $w_{1/2}$. Using first principles of voltammetric theory for a fully reversible reduction, a general expression for $w_{1/2}$ as a function of the applied pulse amplitude (E_{pulse}) and number of electrons (n) involved in the electrochemical reaction (at a fixed temperature) was derived by Parry and Osteryoung.[62] For practical purposes, a fully reversible DP polarogram would exhibit $w_{1/2}$ close to theoretically predicted values. Some common theoretical values of $w_{1/2}$ at 298 K for selected E_{pulse} values (10 – 100 mV) are summarized in Table 6.

Although higher pulses would increase I_{p}, increasing E_{pulse} also increases the width of DPP peaks leading to decrease in resolution. More

Table 6. Theoretical values* of half peak width of differential pulse polarograms involving reversible electrochemical reduction reactions at 298 K

Number of electrons	Pulse amplitude E_{pulse}/mV	Peak width $w_{1/2}$/mV
$n = 1$	10	91
	25	93
	50	99
	75	109
	100	123
$n = 2$	10	46
	25	50
	50	62
	75	80
	100	102
$n = 3$	10	31
	25	36
	50	53
	75	76
	100	100

*Computed using equations in Ref. 62.

importantly, the resultant effect can be used to "test" the interplay of homogeneous and heterogeneous kinetics on recorded peaks.

Values of E_{pulse} between 20 and 100 mV are used in practice as a trade-off between adequate intensities and peak resolution.[61]

Examples of the Gaussian-based curve-fitting of typical DP polarograms (collected for Cd^{II}–glycine–OH system) are presented in Figure 27; extracted $w_{1/2}$ values at different pH stages are plotted in Figure 28. For all practical purposes, the trend in $w_{1/2}$ obtained shows insignificant departure from reversibility; in this instance, $w_{1/2}$ ranged between 61.5 and 66.2 mV and the expected theoretical value for a two-electron reduction reaction probed by DPP (using a pulse amplitude of 50 mV) is 62 mV (see Table 6). This means that the Cd^{II}–glycine–OH system could be regarded as labile and electrochemically reversible; hence, E_p and I_p values obtained from fitting operations (Eqs. 49 and 50) can be directly used by Eq. 17.

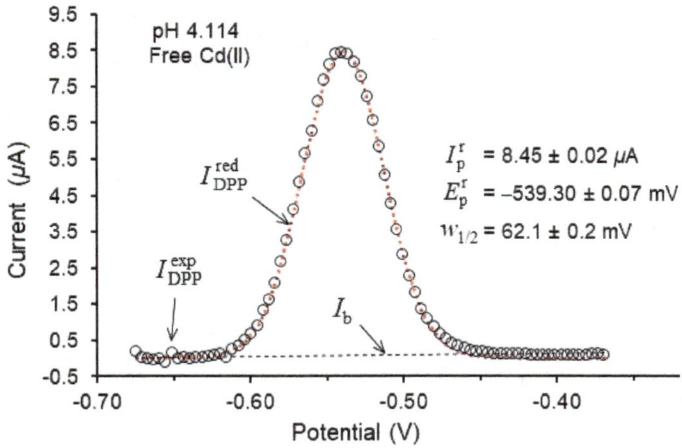

(a)

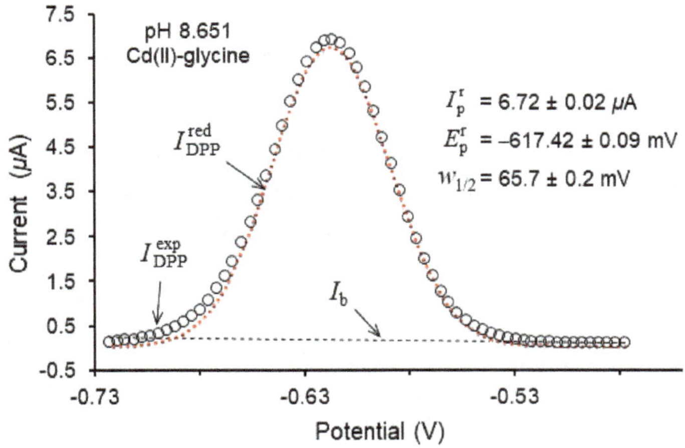

(b)

Figure 27. Examples of fitted DPP peaks recorded at indicated pH values on (a) free Cd(II) and (b) Cd(II)–glycine solution titrated at a fixed $[L]_T$:$[M]_T$ ratio. Open circles (○) are experimental points ($I_{\mathrm{DPP}}^{\mathrm{exp}}$) and dotted lines represent reduction currents ($I_{\mathrm{DPP}}^{\mathrm{red}}$) computed from Eqs. 49 and 50 using refined E_p, I_p, and $w_{1/2}$. DPP parameters: drop time (t_{drop}) = 2 s, pulse time (t_{pulse}) = 200 ms, integration time (t_{int}) = 60 ms, E_{pulse} = 50 mV, E_{step} = 4 mV. Applied potential E is *vs*. Ag/AgCl/3 M KCl reference electrode. Unpublished data.

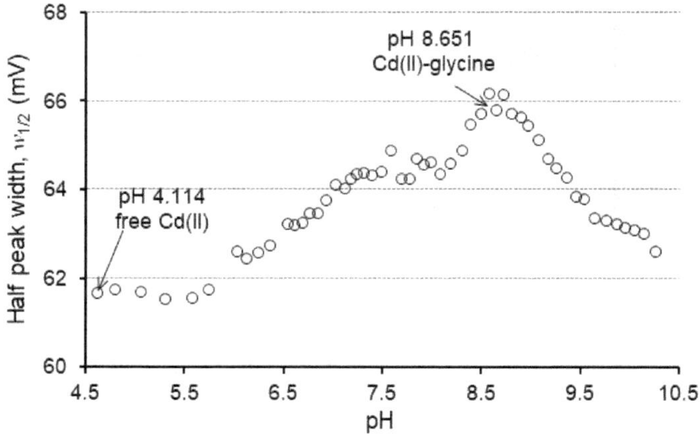

Figure 28. Variation in half peak width, $w_{1/2}$, of DPP peaks for a Cd(II)–glycine–OH system (experimental details as in Figure 27). Unpublished data.

C. Analysis of DC Polarograms

The link between $E_{1/2}$ and the current-potential relationship of a *reversible* reduction process, occurring at the DME, as applied to either conventional DCP or sampled-current DCP, is classically described by the Ilkovič-Heyrovský equation (Eq. 51).[61,63,64]

$$I_{\rm DCP}^{\rm red} = \frac{I_{\rm d}}{\exp\left\{\left(\frac{nF}{RT}\right)\left(E_{\rm step} - E_{1/2}\right)\right\} + 1} \tag{51}$$

For a *totally irreversible* case, the $I_{\rm DCP}^{\rm red}$ *vs.* $E_{\rm appl} = E_{\rm step}$ relationship is described by the extended Ilkovič-Heyrovský equation (Eq. 52),[61,63,64]

$$I_{\rm DCP}^{\rm red} = \frac{I_{\rm d}}{\exp\left\{\left(\alpha\frac{nF}{RT}\right)\left(E_{\rm step} - E_{1/2}\right)\right\} + 1} \tag{52}$$

where $I_{\rm DCP}^{\rm red}$ stands for the DCP reduction current at an applied step potential $E_{\rm step}$, $I_{\rm d}$ is the limiting diffusion current, $E_{1/2}$ is the half-wave potential, α is the cathodic electron transfer coefficient (a dimensionless parameter that may be regarded as a measure of the symmetry of the energy barrier for electron transfer), and n, F, R, and T have their usual meaning.

As the DCP reduction current is described by Eqs. 51 and 52, one can propose a single mathematical function to compute the overall (total) DCP current, Eq. 53:

$$I_{\mathrm{DCP}}^{\mathrm{calc}} = I_{\mathrm{DCP}}^{\mathrm{red}} + I_{\mathrm{b}} = \frac{I_{\mathrm{d}}}{\exp\left\{\left(\delta \dfrac{nF}{RT}\right)\left(E_{\mathrm{step}} - E_{1/2}\right)\right\} + 1} + I_{\mathrm{b}} \qquad (53)$$

where $I_{\mathrm{DCP}}^{\mathrm{calc}}$ stands for the total calculated DCP current at each applied step potential E_{step}, and δ is a reversibility index for DCP. In using the above model, the best approximation of $I_{\mathrm{DCP}}^{\mathrm{red}}$ is obtained by fitting three parameters, $E_{1/2}$, I_{d}, and δ (constrained between 0 and 1) used to reproduce the steepness of the DC wave. For the steepest, *i.e.*, fully reversible, DCP wave one should obtain $\delta = 1$ in accordance with Eq. 51. With departure from reversibility, DCP waves become less steep and the δ parameter becomes smaller than 1. Hence, the δ parameter can be seen as a simple reversibility index. For all practical purposes, δ values above 0.9 indicate electrochemically reversible systems whereas $\delta < 0.9$ indicate departure from reversibility. There is no strict and rigorous criterion in defining reversibility.

It has been established, however, that δ values in the range 0.5 – 0.9 suggest quasi-reversible systems and $\delta < 0.5$ indicate irreversible systems.[32,34,40,41] In a non-linear least squares regression analysis of recorded DCP curves using Eq. 53, the optimized regression parameters are I_{d}, δ, and $E_{1/2}$ (applicable to the DC polarogram) and a, b, c, and d (coefficients of the I_{b} model function). Since polarographic experiments are performed at a fixed temperature and utilize metal ions with stable and known oxidation states, the parameters n, F, R, and T are known constants.

Figure 29 illustrates sampled DC polarograms recorded in a study involving Cd(II) and sarcosine (*N*-methylglycine; $CH_3NH_2CH_2COOH$) at variable pH and fixed $[L]_T$:$[M]_T$ ratio.[41,58] Polarograms were fitted using Eq. 53 in order to extract, besides I_{d}, and observed $E_{1/2}$, δ values needed to assess departures from electrochemical reversibility. Figure 30 illustrates the observed variation in the fitted δ parameter with increasing pH.

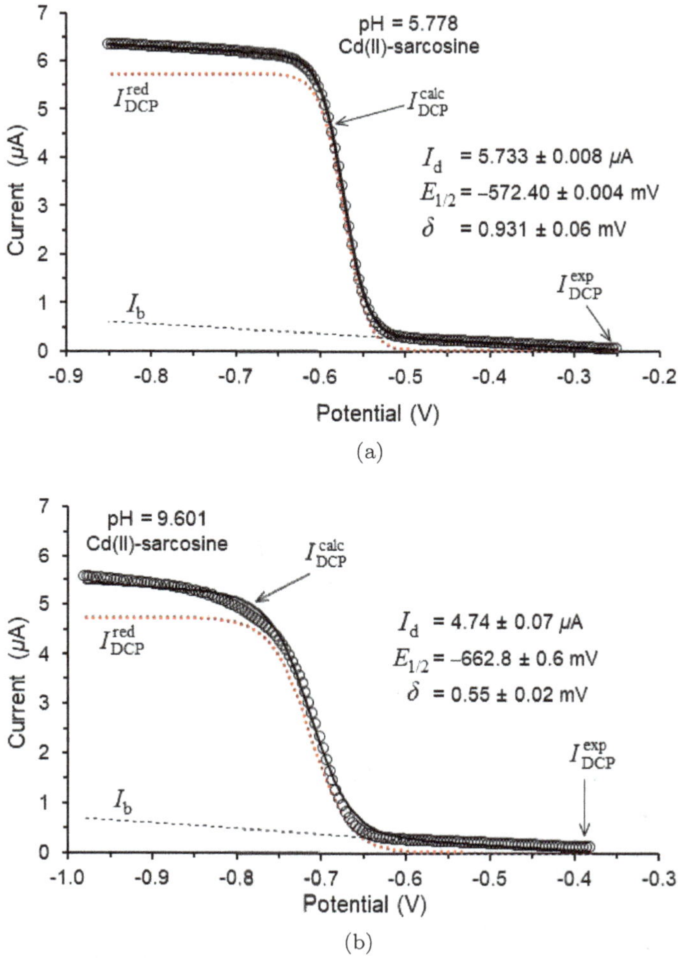

Figure 29. Examples of DC reversible (a) and quasi-reversible (b) polarograms recorded at indicated pH for the Cd(II)-sarcosine-OH system in 0.5 M (H,Na)NO_3 and at 25 °C. Curves were fitted using Eq. 53 with a linear I_b. Applied potential is *vs.* a Ag/AgCl/3 M KCl reference electrode. Initial $[M]_T = 2.040 \times 10^{-4}$ M and $[L]_T = 1.996 \times 10^{-2}$ M. Adapted from Refs. 41 and 58.

To this end, DC polarograms for the free Cd(II) solution and those recorded in the presence of the ligand (up to pH of about 7.3) can be considered as electrochemically reversible. This is because of $\delta > 0.9$ and consequently the fitted $E_{1/2}$ values do not depart significantly (a small fraction of mV) from those obtained for $\delta = 1$. Polarograms recorded

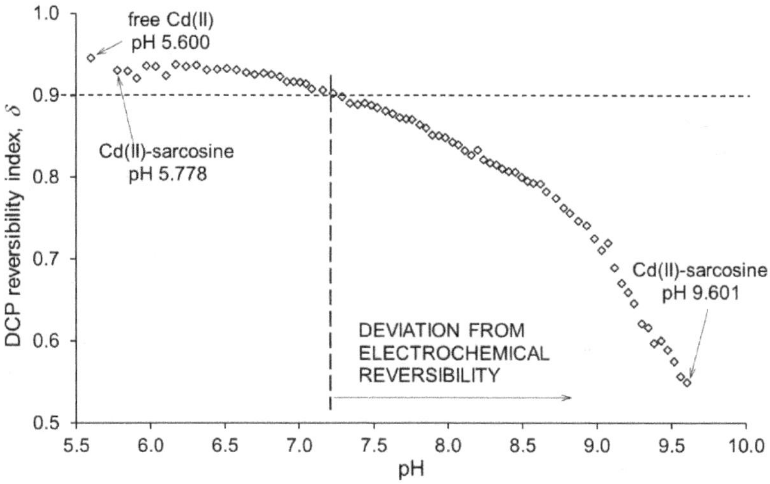

Figure 30. Variation in the DCP reversibility index δ of DC polarograms collected on the Cd^{II}–sarcosine–OH system in a wide pH range. Experimental conditions as in Figure 29. Unpublished data.

above pH 7.3 exhibited quasi-reversible electrochemical reduction characteristics as $0.5 < \delta < 0.9$.

D. Extraction of Reversible Half-wave Potentials from Non-reversible DCP Wave

The curve-fitting model, Eq. 53, can also be adopted to estimate $E^{r}_{1/2}$ values from quasi-reversible experimental DC polarograms. This is practically achieved by fixing δ at unity (as would be expected for a fully reversible DC wave) and manipulating experimental data points between the base of the wave (the predominantly kinetic region) and the polarogram's plateau (the diffusion limiting region). This multistep curve-fitting strategy essentially involves: [32,34]

1. Initial fitting of the experimental polarogram utilizing the full regression model of Eq. 53 in order to obtain I_d, δ, and $E_{1/2}$ and coefficients of I_b (as in Figure 29). Based on the value of δ one can decide if further steps are required. In addition, in some instances it might be necessary to use I_b parameters as fixed values in the following fitting step.

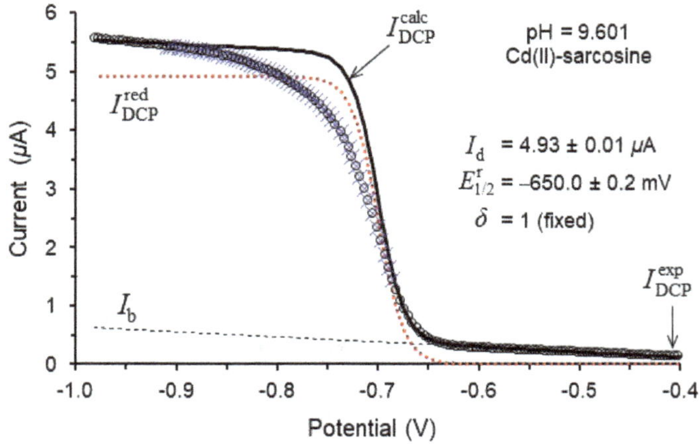

Figure 31. Example of curve-fitting performed on the DC curve (open circles) seen in Figure 29b using Eq. 53. To extract $E_{1/2}^r$ from a quasi-reversible DC polarogram, some points (crosses) were excluded from the fitting procedure. The solid line represents a hypothetical calculated total DCP current for a reversible process, I_{DCP}^{calc}. The dashed line is the fitted background current I_b and the dotted line is the calculated DCP reduction current for a hypothetical fully reversible wave from which $E_{1/2}^r$ and I_d were obtained. Unpublished data.

2. Truncation of the experimental polarogram by exclusion of experimental points above the foot of DC wave (typically above one-fifth of the I_d) as they do not follow the steep rise expected for a reversible case. A sufficient number of points in the limiting diffusion region (the plateau of the DC wave) must be retained for the fitting operation (see Figure 31).
3. Re-fit of the partial set of polarographic experimental points using Eq. 53 with δ fixed at the theoretical value of one.[32] The best-fit $E_{1/2}$ value and I_d thus obtained are taken as $E_{1/2}^r$ and I_d values for subsequent use in computation of complex formation curves and appropriate M–L–OH modelling procedures described in previous sections.

Figure 31 illustrates the analysis of the quasi-reversible DC polarogram, shown in Figure 29b, using the above multistep fitting procedure to extract $E_{1/2}^r$ and I_d values.

Notably, the experimental $E_{1/2}$ value with a δ value of 0.55, Figure 29b, and the estimated $E_{1/2}^r$ with δ fixed at 1 differ by about 13 mV and such a large difference would have a big impact on the refined formation constants. Moreover, one can see that I_d values seen in Figures 29b and 31 differ significantly.

A more versatile curve-fitting strategy for the analysis of non-reversible DC polarograms involves mathematical models derived to fit I–E experimental points of the whole DC polarogram, with $E^{r}_{1/2}$ and I_d one would observe for fully reversible electrochemical process as regression parameters in a non-linear least squares optimization method.[40,41] Matsuda and Ayabe[65–68] derived Eq. 54 for quasi-reversible DC polarographic waves. Importantly, Eq. 54 can reproduce the I–E relationship of a DC quasi-reversible wave by incorporating $E^{r}_{1/2}$ and I_d that one would observe for fully reversible electrochemical process,

$$I^{\text{red}}_{\text{DCP}} = \frac{I_d}{1.13\left(\Lambda\sqrt{\tau}\right)^{-1}\exp\left\{\alpha\frac{nF}{RT}\left(E_{\text{step}} - E^{r}_{1/2}\right)\right\} + \exp\left\{\frac{nF}{RT}\left(E_{\text{step}} - E^{r}_{1/2}\right)\right\} + 1} \tag{54}$$

where the first term, with a heterogeneous kinetic parameter Λ, is of no importance in this case; an interested reader is referred to the original work by Matsuda and Ayabe.[65–68] For the purpose of retrieving the necessary parameters for the M–L studies, *i.e.*, to calculate experimentally recorded quasi-reversible DCP curves, one can make use of Eq. 55 as:

$$\begin{aligned} I^{\text{calc}}_{\text{DCP}} &= I^{\text{red}}_{\text{DCP}} + I_b \\ &= \frac{I_d}{\Gamma\exp\left\{\alpha\frac{nF}{RT}\left(E_{\text{step}} - E^{r}_{1/2}\right)\right\} + \exp\left\{\frac{nF}{RT}\left(E_{\text{step}} - E^{r}_{1/2}\right)\right\} + 1} + I_b \end{aligned} \tag{55}$$

where $\Gamma = 1.13(\Lambda\sqrt{\tau})^{-1}$ is refined in regression analysis, but it does not serve any purpose in the M–L equilibria studies when Eq. 17 is used.

Ružić and co-workers[69] demonstrated that the following relation also holds:

$$1.13\left(\Lambda\sqrt{\tau}\right)^{-1}\exp\left\{\alpha\frac{nF}{RT}\left(E_{\text{step}} - E^{r}_{1/2}\right)\right\} = \exp\left\{\alpha\frac{nF}{RT}\left(E_{\text{step}} - E^{\text{irr}}_{1/2}\right)\right\} \tag{56}$$

where $E_{1/2}^{\mathrm{irr}}$ is a potential corresponding to the irreversible part of the DC wave. The parameter $E_{1/2}^{\mathrm{irr}}$ has no application in the study of metal complexes and, notably, is not equal to the observed from experiment $E_{1/2}$. By incorporating the right-hand term of Eq. 56 into Eq. 54, the reduction current as a function of the applied potential $E_{\mathrm{appl}} = E_{\mathrm{step}}$ can explicitly be described by Eq. 57.

$$I_{\mathrm{DCP}}^{\mathrm{red}} = \frac{I_{\mathrm{d}}}{\exp\left\{\alpha \dfrac{nF}{RT}\left(E_{\mathrm{step}} - E_{1/2}^{\mathrm{irr}}\right)\right\} + \exp\left\{\dfrac{nF}{RT}\left(E_{\mathrm{step}} - E_{1/2}^{\mathrm{r}}\right)\right\} + 1} \tag{57}$$

Hence, a non-linear curve-fitting model (for deconvolution of a non-reversible DC polarograms) can be implemented as[40,41]:

$$I_{\mathrm{DCP}}^{\mathrm{calc}} = \frac{I_{\mathrm{d}}}{\exp\left\{\alpha \dfrac{nF}{RT}\left(E_{\mathrm{step}} - E_{1/2}^{\mathrm{irr}}\right)\right\} + \exp\left\{\dfrac{nF}{RT}\left(E_{\mathrm{step}} - E_{1/2}^{\mathrm{r}}\right)\right\} + 1} + I_{\mathrm{b}} \tag{58}$$

The above curve-fitting model involves I_{d}, $E_{1/2}^{\mathrm{r}}$, α, and $E_{1/2}^{\mathrm{irr}}$ as regression parameters required in calculating the reduction current at a given applied potential. Notably, only the first two parameters are needed in modelling operations of M–L–OH systems.[40,41,58] Figure 32 shows typical examples of reversible and quasi-reversible DC polarograms recorded in a study of the Pb(II)–sarcosine–OH system at various pH values.[40,58]

The recorded DC polarograms in Figure 32 were initially analyzed using Eq. 53 in order to interrogate any variation in electrochemical reversibility (extraction of the δ index and observed $E_{1/2}$ values); observed trends in fitted parameters are presented in Figure 33.[58] The trends in Figure 33 reveal that small differences (< 2 mV) between observed $E_{1/2}$ values (using Eq. 53) and $E_{1/2}^{\mathrm{r}}$ values (retrieved from fitting using Eq. 58) were observed at pH < 7.50. However, large differences of up to 15 mV between the uncorrected $E_{1/2}$ and $E_{1/2}^{\mathrm{r}}$ values were observed at higher pH, where maximum departure from electrochemical reversibility was also observed ($\delta \approx 0.6$ at pH 9.7). It is then seen that the entire set of DC polarograms exhibited diverse electrochemical reversibility as the δ index varied from 0.99 (reversible) to 0.6 (quasi-reversible) between pH 5 and 9.7. Fitting the model based on Eq. 58 allowed for direct evaluation of $E_{1/2}^{\mathrm{r}}$ and

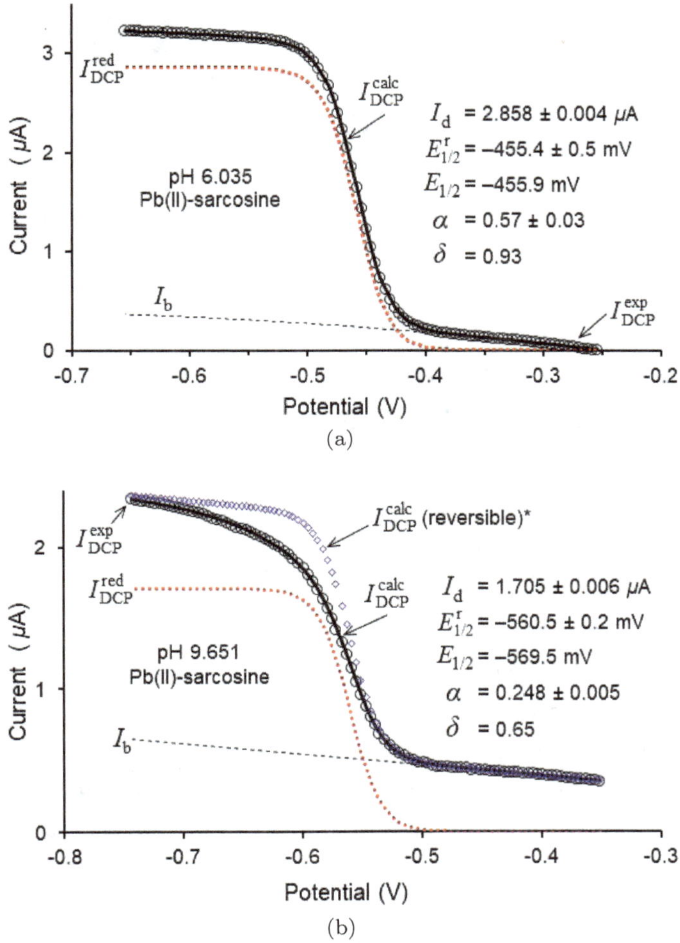

Figure 32. Examples of DC reversible (a) and quasi-reversible (b) polarograms fitted using Eq. 58 (from a study of the Pb^{II}–sarcosine–OH system in 0.5 M $NaNO_3$ at 25 °C, initial $[M]_T = 7.990 \times 10^{-5}$ M, $[L]_T$:$[M]_T$ = 400). For comparison, δ and observed $E_{1/2}$ values were obtained using Eq. 53. The I^{calc}_{DCP} (reversible)* trace stands for the total DCP current corresponding to a hypothetical fully reversible polarogram computed using Eq. 53 with best-fit $E^r_{1/2}$, I_d, and parameters for I_b, all obtained from Eq. 58. Adapted from Refs. 40 and 58.

I_d from the full set of DC polarograms. This made modelling and refinement operations possible and allowed for the determination of a plausible M–L–OH model and reliable log β values for the Pb^{II}–sarcosine–OH system.[40,41,58]

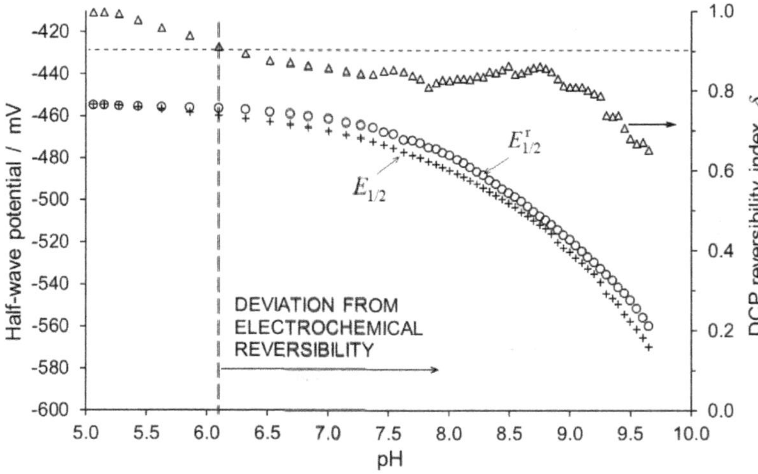

Figure 33. Trends in fitted DCP parameters with increasing pH for the Pb^{II}–sarcosine–OH system. Observed $E_{1/2}$ (pluses) and reversibility index δ (triangles) values were obtained from Eq. 53. $E^{r}_{1/2}$ values (open circles) were computed using Eq. 58. Experimental conditions as in Figure 32. Adapted from Ref. 58.

In rare cases, an insignificant change in the degree of electrochemical reversibility is observed throughout an entire DCP experiment involving reduction of metal complexes over the entire solution composition range of interest (pH or pL). In such cases, to a good approximation, the observed half-wave potential, $E_{1/2}$, can be related to $E^{r}_{1/2}$ by a constant term,[70]

$$E_{1/2} = E^{r}_{1/2} + const \tag{59}$$

Consequently, the experimentally obtained shifts in half-wave potentials $\Delta E_{1/2}$ (relative to $E_{1/2}$(M) obtained for a free metal ion) would essentially be the same as expected for the hypothetical reversible potentials over the entire range of measurements; that is,

$$\Delta E_{1/2} = \Delta E^{r}_{1/2} \tag{60}$$

In such instances, the observed $E_{1/2}$ from DCP (which can be simply extracted using fits involving Eq. 53) can directly be used in the derivation of plausible models using methodologies described in this chapter and in more detail elsewhere.[37]

From the above it follows that DPP, being more sensitive, should preferentially be used provided that (i) peaks do represent reversible processes or (ii) departure from reversibility is within specific limits and does not vary significantly throughout the experimental range of solution compositions. Otherwise, DCP is the preferred mode in studying polarographically non-reversible M–L systems.

E. Treatment of Overlapping Polarographic Signals

In resolving two or more overlapped DPP peaks (or DCP waves) by curve fitting, the main assumption is that the electrochemical processes generating the observed current proceed independently with closely spaced peak potentials or half-wave potentials. Treatment of such signals generally requires a constructed current–potential model to account for each conceivable electrochemical process contributing to the total reduction current. For example, a simple strategy for deconvoluting overlapping DPP peaks by multivariate non-linear curve fitting[29] involves summing modified Gaussian equations (Eq. 61) that mathematically describe the variation of individual DPP reduction current at any applied E,

$$\begin{aligned} I_{\text{DPP}}^{\text{calc}} &= I_{\text{DPP}}^{\text{red(1)}} + I_{\text{DPP}}^{\text{red(2)}} + \cdots + I_{\text{DPP}}^{\text{red}(n)} + I_{\text{b}} \\ &= \sum_{i=1}^{n} I_{\text{p}(i)} \exp\left\{ -\left(E_{\text{step}} - E_{\text{p}(i)}\right)^2 \Big/ 2\left(w_{1/2(i)}/2.345\right)^2 \right\} + I_{\text{b}} \end{aligned} \quad (61)$$

where $I_{\text{DPP}}^{\text{calc}}$ is the total calculated DPP current, $I_{\text{DPP}}^{\text{red}(i)}$ is the DPP reduction current for the ith species, $E_{\text{p}(i)}$ is the peak potential corresponding to the ith peak, $I_{\text{p}(i)}$ is the peak current of ith peak, and $w_{1/2(i)}$ is the half-peak width corresponding to the ith peak.

Similarly, overlapping DC waves involving two or more electrochemical reduction processes can be calculated by summing up equations relating individual DCP reduction currents as a function of applied E and appropriate regression parameters refined *via* the usual non-linear least squares optimization procedures.[29]

XIII. General Conclusions

Although polarographic studies of metal–ligand equilibria goes back about 80 years (Lingane was first to report in 1941 relevant mathematical expressions for fully labile metal–ligand systems studied at fixed pH and large $[L]_T:[M]_T$ ratios[1]) the majority of reported formation constants comes from glass electrode potentiometry (GEP). This means that for decades, polarography was not an analytical method of choice and this is due to many limitations of the polarographic protocols reported. For instance, pH-titrations were not considered in polarographic studies at all (titration of a sample solution with a concentrated ligand solution was the main practice for decades) and mass-balance equations were not incorporated in experimental data treatment.

Dominance of GEP is not surprising as it can be used to study M–L equilibria involving any metal ion and ligand that form homogeneous solution in the wide pH range, typically between 2 and 12. Moreover, GEP is perfect for determination of protonation constants of ligands. GEP, as any other analytical technique, however, has severe limitations that can be seen as a window of opportunity for other analytical techniques. Considering polarography, a paradigm shift took place in 1996 (more than a half a century after the first Lingane's report) when a new expression was reported in the polarographic studies of metal–ligand equilibria that was developed for essentially the same approach as implemented in GEP. Notably, polarographic and potentiometric experiments became compatible in that they could be performed as a combined experiment on the same sample solution at the fixed $[L]_T:[M]_T$ ratios and varying pH. In fact, to execute a successful polarographic experiment, the glass electrode should be calibrated using most rigorous protocols implemented in GEP. pH readings are taken to ±0.1 mV throughout entire experiment in order to compute the free proton concentration at each titration point. The computed free proton concentrations are then used in solving mass-balance equations, written for the assumed M–L–OH model, from which the free metal ion concentration is obtained and subsequently used to reproduce the experimentally recorded data in the form of the corrected shift (Eq.18). The M–L–OH model is fine-tuned and formation constants refined such that computed complex formation curve fits the experimental one best.

Importantly, the same mathematics used in refinement of formation constants is also employed in prior-experiment modelling procedures – from which one might predict major species expected to form in a solution (without any consideration as to whether the M–L system is fully labile or inert). This is possible because the same general expression (Eq. 17) is valid in the study of labile, non-labile (inert) and kinetically mixed metal–ligand systems, provided that polarographic signals recorded on a labile part of the system are used in modelling and refinement operations. To this effect, numerous protocols are presented in this chapter that explain in detail how to treat polarographic data with a rigor needed to produce reliable outputs, *i.e.*, models and formation constants.

Finally, additional theory described herein, and relevant (simple) mathematical expressions, show how to transform polarographic data in the potentiometric format that is suitable for refinement operations generally performed by software dedicated to potentiometry. In general, the protocols described should be seen as a complete package that can be used through an entire process applicable to the study of metal–ligand equilibria by polarographic techniques and this includes:

1. Pre-experiment modelling (to predict best experimental conditions, such as $[M]_T$, $[L]_T$, $[L]_T$:$[M]_T$ ratio, pH range, *etc.*),
2. Recommendation on how to execute experiments successfully.
3. Fitting experimental polarographic DCP and DPP curves to "extract" most accurate reversible half-wave and peak potentials as well as signal intensities.
4. Identifying major complexes formed by analysis of slopes of experimental complex formation curves.
5. Strategies that are useful in fitting computed formation curves into experimental complex formation curves (from which final model and refined formation constants are generated).
6. Planning consecutive experiments that might prove most useful in confirming the proposed model or improving formation constants of minor species.

Hence, provided that a metal ion of interest is (i) polarographically active in available applied potential window and (ii) interpretable and generates a measurable signal in the selected background electrolyte when a ligand of interest is present, it should be possible to determine high quality and reliable M–L–OH models (and rigorously refined formation constants) when described protocols are followed, even when many complexes (polynuclear, protonated, hydroxo complexes, *etc.*) are formed simultaneously in a solution sample.

XIV. References

1. Lingane, J. J. Interpretation of the Polarographic Waves of Complex Metal Ions. *Chem. Rev.* **1941**, *29*, 1–35.
2. DeFord, D. D.; Hume, D. N. The Determination of Consecutive Formation Constants of Complex Ions from Polarographic Data. *J. Am. Chem. Soc.* **1951**, *73*, 5321–5322.
3. Hume, D. N.; DeFord D. D.; Cave, G. C. B. A Polarographic Study of the Cadmium Thiocyanate Complexe. *J. Am. Chem. Soc.* **1951**, *73*, 5323–5325.
4. Schaap, W. B.; McMasters, D. L. A Polarographic Study of Mixed-ligand Complex Formation; Complexes of Copper and Cadmium with Oxalate Ion and Ethylenediamine. *J. Am. Chem. Soc.* **1961**, *83*, 4699–4706.
5. Crow, D. R. *Polarography of Metal Complexes*; Academic Press: London and New York, 1969.
6. Koryta, J. *Principles in Polarography*; Heyrovský, J. and Kûta, J. Eds.; Academic Press: New York, 1966.
7. Anderegg, G.; Nägeli, P.; Müller, F.; Schwarzenbach, G.. Komplexone XXX. Diäthylentriamin-pentaessigsäure (DTPA). *Helv. Chim. Acta* **1959**, *42*, 827–836.
8. Schwarzenbach, G.; Gut, R.; Anderegg, G. Komplexone XXV. Die polarographische Untersuchung von Austauschgleichgewichten. Neue Daten der Bildungskonstanten von Metallkomplexen der Äthylendiamin-tetraessigsäure und der 1,2-Diaminocyclohexan-tetraessigsäure. *Helv. Chim. Acta* **1954**, *37*, 937–957.
9. De Jong, H. G.; van Leeuwen, H. P.; Holub. K. Voltammetry of Metal Complex Systems with Different Diffusion Coefficients of the Species Involved. PART I. Analytical Approaches to the Limiting Current for the General Case Including Association/Dissociation Kinetics. *J. Electroanal. Chem.* **1987**, *234*, 1–16.
10. Díaz-Cruz, J. M.; Ariño, C.; Esteban, M.; Casassas, E. Polarography and Anodic Stripping Voltammetry of Metalpolycarboxylate Complexes: Phenomenological Relationship between Limiting Currents and Hydrodynamic Mass Transport. *J. Electroanal. Chem.* **1992**, *333*, 33–45.
11. Martell, A. E.; Smith, R. M. *Critical Stability Constants*; Plenum Press; New York, 1974–1989, Vol. 1–6.
12. Smith, R. M.; Martell, A. E. *NIST Standard Reference Database 46. NIST Critically Selected Stability Constants of Metal Complexes Database. Version 3.0.* Data

collected and selected by R. M. Smith and A. E. Martell. U.S. Department of Commerce, National Institute of Standards and Technology, 1997.

13. Martell, A. E.; Hancock, R. D.; *Metal Complexes in Aqueous Solutions*; Modern Inorganic Chemistry series, Fackler, J. P., Ed.; Plenum Press: New York, 1996.
14. Hancock, R. D.; Cukrowski, I.; Baloyi, J.; Mashishi, J. The Affinity of Bismuth(III) for Nitrogen-Donor Ligands. *J. Chem. Soc., Dalton Trans.* **1993**, 2895–2899.
15. Hancock, R. D.; Cukrowski, I.; Antunes, I.; Cukrowska, E.; Mashishi, J.; Brown, K. Complexation of $Bismuth^{III}$ by Nitrogen Donor Ligands. A Polarographic Study. *Polyhedron* **1995**, *14*, 1699–1707.
16. Cukrowski, I.; Cukrowska, E.; Hancock, R. D.; Anderegg, G. The Effect of Chelate Ring Size on Metal Ion Size-based Selectivity in Polyamine Ligands Containing Pyridyl and Saturated Nitrogen Donor Groups. *Anal. Chim. Acta* **1995**, 312, 307–321.
17. Cukrowski, I.; Marsicano, F.; Hancock, R. D.; Tshetlho, P. T.; van Otterlo, W. A. L. The Neutral Oxygen Donor in Complexes of Lead and Cadmium. A Potentiometric, Calorimetric and DP Polarographic Study. *Polyhedron* **1995**, *14*, 1661–1674.
18. Cukrowski, I.; Hancock, R. D.; Luckay, R. C. Formation Constant Calculation for Non-Labile Complexes Based on a Labile Part of the Metal–Ligand System. A Differential Pulse Polarographic Study at Fixed Ligand to Metal Ratio and Varied pH: Application to Polarographically Inactive Complexes. *Anal. Chim. Acta* **1996**, *319*, 39–48.
19. Cukrowski, I. A Polarographic Method of Speciation for Labile Metal–Ligand Systems Based on Mass–Balance Equations. A Differential Pulse Polarographic Study at Fixed Ligand to Metal Ratio and Varied pH. *Anal. Chim. Acta* **1996**, *336*, 23–36.
20. Cukrowski, I.; Adsetts, M. Experimental and Calculated Complex Formation Curves for a Labile Metal–Ligand System. A Differential Pulse Polarographic Study of Pb(II)–(*N,N,N′,N′*-tetramethylethylenediamine)–OH System at Fixed Ligand to Metal Ratio and Varied pH. *J. Electroanal. Chem.* **1997**, *429*, 129–137.
21. Cukrowski, I. Modelling of Cd^{II} Species Formed with *N,N′*-bis(2-pyridylmethyl)-1, 2-diaminoethane and Optimisation of Stability Constants by the Use of Polarographic Complex Formation Curves for Labile Metal–Ligand Systems. *Analyst* **1997**, *122*, 827–833.
22. Cukrowski, I. Experimental and Calculated Complex Formation Curves as an Optimisation Tool in Speciation Studies of Polarographically Labile Metal–Ligand Systems. *Electroanalysis* **1997**, *9*, 699–706.
23. Cukrowski, I. Modelling of a Cd^{II}–N-(2-hydrohyethyl)ethylenediamine System and Optimisation of Stability Constants by the Use of Complex Formation Curves. A Polarographic Study at a Fixed L:M Ratio and Varied pH. *Electroanalysis* **1997**, *9*, 1167–1173.
24. Cukrowski I.; Loader, S. A. Evaluation of a Peak Potential of an Uncomplexed Metal Ion in Speciation Study of Labile Metal–Ligand Systems by Polarography. Stability

Constants Determination of Cd^{II} and Pb^{II} with Picolinic Acid at a Fixed L_T:M_T Ratio and Varied pH. *Electroanalysis* **1998**, *10*, 877–885.

25. Cukrowski, I. Experimental and Calculated Complex Formation Curves for Mixed, Dynamic and Semi-Dynamic, Metal–Ligand Systems. A Differential Pulse Polarographic Study of Cu^{II}–Sarcosine–OH and Cd^{II}–MIDA–OH Systems at a Fixed Ligand to Metal Ratio and Varied pH. *J. Electroanal. Chem.* **1999**, *460*, 197–206.
26. Cukrowski, I.; Zeevaart, J. R.; Jarvis, N. V. A Potentiometric and Differential Pulse Polarographic Study of Cd^{II} with 1-Hydrohyethylenediphosphonic Acid. *Anal. Chim. Acta* **1999**, *379*, 217–226.
27. Cukrowski, I. Polarographic Complex Formation Curves for Fully Inert Metal–Ligand System. *Electroanalysis* **1999**, *11*, 606–613.
28. Cukrowski, I.; Ngigi, G. Solution Equilibria. A Unified Mathematical Treatment of Potentiometric and Polarographic Data in a Metal–Ligand Equilibrium Study at a Fixed Ligand to Metal Ratio and Various pH Values. *Electroanalysis* **2001**, *13*, 1242–1252.
29. Cukrowski, I.; Maseko, N. Solution Equilibria. A Unified Mathematical Treatment of Experimental Polarographic and Potentiometric Data from Acid–Base and Ligand Titrations. A Polarographic and Ion Selective Electrode Study of Cd^{II}–(N-2-hydroxyethyl)imino-diacetic acid)–OH System. *Electroanalysis* **2003**, *15*, 1377–1388.
30. Machado, C. M. M.; Cukrowski, I.; Gameiro, P.; Soares, H. M. V. M. Challenges in Modelling and Optimisation of Stability Constants in the Study of Metal Complexes with Monoprotonated Ligands. A Glass Electrode Potentiometric and Polarographic Study of a Cu–TAPSO–OH System. *Anal. Chim. Acta* **2003**, *493*, 105–119.
31. Machado, C. M. M.; Cukrowski, I.; Soares, H. M. V. M. Challenges in Modelling and Optimisation of Stability Constants in the Study of Metal Complexes with Monoprotonated Ligands. Part II: A Glass Electrode Potentiometric and Polarographic Study of Cu–AMPSO–OH System. *Helv. Chim. Acta* **2003**, *86*, 3288–3304.
32. Cukrowski, I.; Zhang, J. M. Voltammetry as a Virtual Potentiometric Sensor in Modelling of a Metal–Ligand System and Refinement of Stability Constants. Part 1. Polarographic and Glass Electrode Potentiometric Study of a Dynamic Cd–Glycine System. *Electroanalysis* **2004**, *16*, 612–626.
33. Machado, C. M. M.; Scheerlinck, S.; Cukrowski, I.; Soares, H. M. V. M. Challenges in Modelling and Optimisation of Stability Constants in the Study of Metal Complexes with Monoprotonated Ligands. Part III: A Glass Electrode Potentiometric and Polarographic Study of Cu–DIPSO–OH System. *Anal. Chim. Acta* **2004**, *518*, 117–126.
34. Cukrowski, I.; Zhang, J. M.; van Aswegen, A. Voltammetry as a Virtual Potentiometric Sensor in Modelling of a Metal/Ligand System and Refinement of Stability Constants. Part 2. Differential-Pulse and Sampled-Current-Polarographic and Virtual Free-Metal-Ion Potentiometric Study of a Bismuth(III)/Picolinic Acid/Hydroxide System. *Helv. Chim. Acta* **2004**, *87*, 2135–2158.

35. Cukrowski, I.; Zhang, J. M. Voltammetry as a Virtual Potentiometric Sensor in Modelling of a Metal–Ligand System and Refinement of Stability Constants. Part 3. Differential-Pulse- and Sampled-Current-Polarographic and Virtual Free Metal Ion Potentiometric Study of a Bismuth(III)–EDDA–Hydroxide System. *Chem. Anal.* **2005**, *50*, 3–36.
36. Machado, C. M. M.; Cukrowski, I.; Soares, H. M. V. M. Graphic Data Analysis and Complex Formation Curves as Modelling and Optimisation Tools in Potentiometric and Voltammetric Speciation Studies of a Pb–$(TAPSO)_x$–$(OH)_y$ System. *Electroanalysis* **2005**, *17*, 1291–1301.
37. Cukrowski, I.; Mogano, D. M.; Zeevaart, J. R. Voltammetry as a Virtual Potentiometric Sensor in Modelling of a Metal–Ligand System and Refinement of Stability Constants. Part 4. An Electrochemical Study of Ni^{II} complexes with Methylene Diphosphonic Acid (MDP). *J. Inorg. Biochem.* **2005**, *99*, 2308–2321.
38. Machado, C. M. M.; Cukrowski, I.; Soares, H. M. V. M. Interpretation of Non-Nernstian Slopes in Graphic Analysis of Data Collected in pH-Range Close to Deprotonation of a Ligand. Part I. A Glass Electrode Potentiometric and Polarographic Study of Cd–$(TAPSO)_x$–$(OH)_y$ and Zn–$(TAPSO)_x$–$(OH)_y$ Systems. *Talanta* **2006**, 68, 819–830.
39. Machado, C. M. M.; Cukrowski, I.; Soares, H. M. V. M. Complex Formation in the Region of Metal Hydrolysis and $M(OH)_2$ Precipitation. A Glass Electrode Potentiometric and Polarographic Study of Cd–$(AMPSO)_x$–$(OH)_y$ and Zn–$(AMPSO)_x$–$(OH)_y$ Systems. *Electroanalysis* **2006**, *18*, 719–729.
40. Cukrowski, I.; Magampa, P.; Mkwizu, T. S. Voltammetry as Virtual Potentiometric Sensor in Modelling of a Metal/Ligand System and Refinement of Stability Constants. Part 5. Complexation Studies of Hydrolysis-Prone Lead(II) with Glycine and Sarcosine by Sampled-Direct-Current Polarography Involving Virtual Potential. *Helv. Chim. Acta* **2006**, *89*, 2934–2952.
41. Cukrowski, I.; Marques, H. M.; Mkwizu, T.; Magampa, P. P.; Serge, C. Influence of Electronic and Steric Effects on Stability Constants and Electrochemical Reversibility of Divalent ion Complexes with Glycine and Sarcosine. A Glass Electrode Potentiometric, Sampled Direct Current Polarographic, Virtual Potentiometric, and Molecular Modelling Study. *Anal. Chim. Acta* **2007**, *590*, 203–216.
42. Billing, C.; Cukrowski, I. Glass Electrode Calibration for Use in the Voltammetric Determination of Stability Constants under Extreme Acidic Conditions. *S. Afr. J. Chem.* **2009**, *62*, 168–178.
43. Billing, C.; Cukrowski, I.; Jordan, B. A Novel Approach to Monitoring of the Diffusion Junction Potential in Speciation Studies by. Polarography Under very Acidic Conditions. Part I: The Reversible Cd(II)-Picolinic Acid System. *Electroanalysis* **2013**, *25*, 2221–2230.
44. Billing, C.; Cukrowski, I. A Novel Approach to Monitoring of the Diffusion Junction Potential in Speciation Studies by Polarography Under very Acidic Conditions. Part II: The quasi-reversible Cu(II)–Picolinic Acid System. *Electroanalysis* **2015**, *27*, 494–502.

45. Billing, C.; Cukrowski, I. Measurements and Modelling to Determine the Reduction Potential of Uncomplexed Bi(III) in Nitrate Solutions for Application in Bi(III)–Ligand Equilibria Studies by Voltammetry. *J. Phys. Chem. B* **2016**, *120*, 4268–4278.
46. Billing, C.; Cukrowski, I. Application of Protocols Devised to Study Bi(III) Complex Formation by Voltammetry: The Bi(III)–Picolinic Acid System. *J. Phys. Chem. B* **2016**, *120*, 12972–12980.
47. May, P. M.; Williams, D. R.; Linder, P. W.; Torrington, R. G. The Use of Glass Electrodes for the Determination of Formation Constants — I. A Definitive Method for Calibration. *Talanta* **1982**, *29*, 249–256.
48. May, P. M.; Murray, K.; Williams, D. R. The Use of Glass Electrodes for the Determination of Formation Constants — II. Simulation of Titration Data. *Talanta* **1985**, *32*, 483–489.
49. May, P. M.; Murray, K.; Williams, D. R. The Use of Glass Electrodes for the Determination of Formation Constants — III. Optimization of Titration Data: The ESTA Library of Computer Programs. *Talanta* **1988**, *35*, 825–830.
50. May, P. M.; Murray, K. The Use of Glass Electrodes for the Determination of Formation Constants — IV Matters of Weight. *Talanta* **1988**, *35*, 927–932.
51. Bates, R. G. *Determination of pH. Theory and Practice*. 2nd ed.; John Wiley and Sons: New York, 1973.
52. Irving, H. M.; Miles, M. G.; Pettit, L. D. A Study of Some Problems in Determining the Stoichiometric Proton Dissociation Constants of Complexes by Potentiometric Titrations Using a Glass Electrode. *Anal. Chim. Acta* **1967**, *38*, 475–488.
53 Nancollas, G. H.; Tomson, M. B. Guidelines for the Determination of Stability Constants. *Pure Appl. Chem.* **1982**, *54*, 2675–2692.
54 Brown, P. L.; Ekberg, C. *Hydrolysis of Metal Ions*; Wiley-VCH: Weinheim, Germany, 2016.
55. Fiol, S.; Arce, F.; Armesto, X. L.; Penedo, F.; Sastre de Vicente, M. Analysis of Systematic Errors in Calibrating Glass Electrodes with H^+ as a Concentration Probe. *Fresenius J. Anal. Chem.* **1992**, *343*, 469–472.
56. Hitchcock, D. I.; Taylor, A. E. The Standardization of Hydrogen Ion Determinations. I. Hydrogen Electrode Measurements with a Liquid Junction. *J. Am. Chem. Soc.* **1937**, *59*, 1812–1818.
57. Ngigi, G. K. Metal–Ligand Equilibrium Studies by Polarographic Techniques and Ion Selective Electrodes. M.Sc. Dissertation, University of the Witwatersrand, Johannesburg, South Africa, 2001.
58. Mkwizu, T. S. P. A Polarographic and Potentiometric Study of Metal–Ligand Equilibria: Instrumentation and Investigations of Systems with Non-Reversible Electrode Reactions. M.Sc. Dissertation, University of the Witwatersrand, Johannesburg, South Africa, 2006.
59. Huang, W.; Henderson, T. L. E.; Bond, A. M.; Oldham, K. B. Curve Fitting to Resolve Overlapping Voltammetric Peaks: Model and Examples. *Anal. Chim. Acta* **1995**, *304*, 1–15.

60. Rusling, J. F.; Kumosinski, T. F. *Nonlinear Computer Modelling of Chemical and Biochemical Data*; Academic Press: London, 1996.
61. Bard, A. J.; Faulkner, L. R. *Electrochemical Methods: Fundamentals and Applications*; John Wiley & Sons: New York, 2001.
62. Parry, E. P.; Osteryoung, R. A. Evaluation of Analytical Pulse Polarography. *Anal. Chem.* **1965**, *37*, 1634–1637.
63. Bond, A. M. *Modern Polarographic Methods in Analytical Chemistry*; Marcel Dekker: New York, 1980.
64. Heyrovsky, J.; Kuta, J. *Principles of Polarography*; Academic Press: Prague, 1966.
65. Matsuda, H.; Ayabe, Y. Polarographische Untersuchungen über die Kinetik der Entladung von Zink-Tartrat-Komplexen (Polarographic Studies on the Kinetics of the Discharge of Zinc Tartrate Complexes). *Z. Elektrochem.* **1962**, *66*, 469–477.
66. Matsuda, H.; Ayabe, Y. Polarographische Untersuchungen über die Kinetik der Entladung von Komplex-Metallionen, insbesondere von Hydroxo und Ammin-komplexen des Zinks (Polarographic Investigations into the Discharge Kinetics of Complex Metal Cations, namely Hydroxo and Amine Complexes of Zinc). *Z. Elektrochem.* **1959**, *63*, 1164–1171.
67. Matsuda, H.; Ayabe, Y. Theoretical Analysis of Polarographic Waves. II. Reduction of Complex Metal Ions. *Bull. Chem. Soc. Jpn.* **1956**, *29*, 134–140.
68. Matsuda, H.; Ayabe, Y. Theoretical Analysis of Polarographic Waves. I. Reduction of Simple Metal Ions. *Bull. Chem. Soc. Jpn.* **1955**, *28*, 422–428.
69. Ružić, I.; Barić, A.; Branica, M. Determination of Electrode Reaction Parameters ($E^{r}_{1/2}$ and αn) of the Quasi-Reversible D.C. polarographic waves. *J. Electroanal. Chem.* **1971**, *29*, 411–416.
70. Bond, A. M. A Proposed Simple Method of Calculation of Stability Constants for Some Non-Reversible Electrode Reactions at the Dropping Mercury Electrode. *J. Electroanal. Chem.* **1970**, *28*, 433–442.

https://doi.org/10.1142/9789811281099_0002

2 AGNES, an Equilibrium Technique for the Robust Speciation of Metals and Metalloids

Josep Galceran,* Encarna Companys, Calin A. David, Carlos Rey-Castro and Jaume Puy

Dep. Chemistry, Physics and Environmental and Soils Sciences. Universitat de Lleida, and AGROTECNIO-CERCA, Rovira Roure 191, 25198 Lleida, Catalonia, Spain
*josep.galceran@UdL.cat

Table of Contents

I. List of Abbreviations

AGNES	absence of gradients and Nernstian equilibrium stripping
AGNES-1P	AGNES one pulse
AGNES-2P	AGNES two pulses
AGNES-I	AGNES using current from stripping stage
AGNES-Q	AGNES using charge from stripping stage
AGNELSV	AGNES with linear stripping voltammetry
AGNES-SCP	AGNES with chronopotentiometry
ASV	anodic stripping voltammetry
BiFE	bismuth film electrode
CAS	conditional affinity spectrum
CLE-AdCSV	competitive ligand exchange adsorptive cathodic stripping voltammetry
DGT	diffusion gradients in thin-films
DMEM	Dulbecco's modified Eagle medium
DMT	Donnan membrane technique
DOM	dissolved organic matter
DPP	differential pulse polarography
EDTA	ethylenediaminetetraacetic acid
GSH	glutathione
HMDE	hanging mercury drop electrode
IDA	iminodiacetic acid
ICP-MS	inductively coupled plasma mass spectrometry
IET	ion exchange technique
Ir-μE	iridium plated microelectrode
ISE	ion selective electrode
LOD	limit of detection
NM	nanomaterial

NICA	non-ideal competitive adsorption
NTA	nitrilotriacetic acid
PAA	polyacrylic acid
PDCA	pyridinedicarboxylic acid
PIM	polymer inclusion membrane devices
PLM	permeation liquid membrane
RDE	rotating disc electrode
RT	resin titration
SI	supporting information
SSCP	scanned stripping chronopotentiometry
SPE	screen-printed electrode
TMF-RDE	thin mercury film rotating disc electrode
VGME	vibrating gold microwire electrode
VM	Visual MINTEQ
WHAM	Windermere humic acid model

II. List of Symbols

A. Latin Symbols

Symbol	Description	Units
A	electrode surface area	m^2
$c_{T,X}$	total concentration of species X in the solution	$mol\ L^{-1}$ or $mol\ m^{-3}$
c_{ZnO}	total mass concentration of ZnO in the dispersion	$kg\ m^{-3}$
$D_{M^{n+}}$	diffusion coefficient of ion M^{n+} in solution	$m^2\ s^{-1}$
D_{M^0}	diffusion coefficient of M^0 (reduced metal) in the amalgam	$m^2\ s^{-1}$

Symbol	Description	Units
D_{ML}	diffusion coefficient of complex ML in the solution	$m^2\ s^{-1}$
D_X	diffusion coefficient of species X in solution	$m^2\ s^{-1}$
E_j	potential associated to a given gain	V
$E^{0'}$	formal standard potential of the redox couple	V
E_1	deposition potential (associated to the gain Y)	V
$E_{1,a}$	deposition potential under diffusion-limited conditions (in the first sub-stage of AGNES-2P)	V
E_2	stripping potential (in AGNES-I or AGNES-Q)	V
EC	electrical conductivity	$\mu S\ cm^{-1}$
E_{calib}	deposition potential applied in the calibration	V
E_{peak}	DPP peak potential	V
F	Faraday constant	$C\ mol^{-1}$
I	faradaic intensity current	A
I_{lim}	average current at the end (*e.g.* just before t_2 = 50 s) of the second stage of AGNES-I or AGNES-Q	A
I_{meas}	measured current	A
I_{Ox}	oxidants current (in the resting period at the end of the deposition stage)	A
I_s	prescribed stripping current in AGNES-SCP	A
J	flux	$mol\ m^{-2}\ s^{-1}$
K_{sp}	solubility product	various
K_X^{cond}	conditional formation constant for the corresponding specie X	various
K'	overall conditional formation constant corresponding to the medium composition	none

Symbol	Description	Units
L	vacant macromolecular site or free ligand	none
M	metal	none
M^{n+}	free metal (hexa-aquo ion)	none
M^0	reduced metal	none
ML	complexed metal in solution	none
$[M^{n+}]$	free metal concentration	mol L^{-1} or mol m^{-3}
$[M^0]$	concentration of reduced metal inside the amalgam	mol L^{-1} or mol m^{-3}
[ML]	complexed metal concentration	mol L^{-1} or mol m^{-3}
n	number of exchanged electrons	none
Q	faradaic charge	C
r	radius of the primary particles	m
r_0	radius of the mercury drop in HMDE radius of an aggregate of nanoparticles	m
R	gas constant	J K^{-1} mol^{-1}
S	solubility	mol L^{-1} or mol m^{-3} or g L^{-1}
T	temperature	K
t_1	deposition time in AGNES-1P	s
$t_{1,a}$	deposition time of the first sub-stage in AGNES-2P	s
$t_{1,b}$	deposition time of the second sub-stage in AGNES 2P	s
t_2	time of the second stage or reoxidation time	s
$t_{1/2}$	time necessary to reach a free Zn^{2+} concentration of 0.5 times the concentration at equilibrium	s

Symbol	Description	Units
t_w	waiting time (with no stirring) at the end of the first stage	s
V_{Hg}	volume of the mercury electrode	L
$\overline{V}$	molar volume of bulk $ZnO_{s,zincite}$	$m^3\ mol^{-1}$
Y	gain or preconcentration factor achieved by the end of the first stage (also Y_1 in some literature)	none
Y_{calib}	gain or preconcentration factor corresponding to E_{calib}	none
Y_j	gain or preconcentration factor corresponding to E_j	none
Y'	gain for a non-free concentration (*i.e.* concentration of species other than the free metal ion), *e.g.* for $Sb(OH)_3$	none

B. Greek Symbols

Symbol	Description	Units
γ	$ZnO_{s,zincite}$/water interfacial energy	$J\ m^{-2}$
γ_X	activity coefficient of species X	none
δ	effective diffusion layer thickness	m
ΔE	modulation amplitude	V
$\Delta H°$	standard enthalpy of dissolution	$J\ mol^{-1}$
η	proportionality factor between the faradaic stripped current (I) and the reduced metal concentration	$A\ L\ mol^{-1}$
η_Q	proportionality factor between the faradaic stripped charge (Q) and the reduced metal concentration	$C\ L\ mol^{-1}$
μ	ionic strength	$mol\ L^{-1}$
ξ	lability degree of a complex ML (or pool of complexes)	none
τ	transition time (in the SCP measurement)	s

III. Introduction

Analytical techniques for chemical speciation, such as those dealt with in this book, can be classified as equilibrium or dynamic techniques.

Equilibrium techniques are those that report back a well-defined equilibrium parameter of the system (with respect to some process in which we are interested). Since this chapter focusses just on metal (or metalloid) complexation (with a fixed redox state of the element), the typical reported information in an equilibrium technique is the free metal ion concentration. Amongst the equilibrium techniques, without being exhaustive, one can mention: (1) Ion Selective Electrodes (ISE, including pH potentiometry),[1] based on measuring a potential difference calibrated in terms of an ion activity (or concentration); (2) Ion Exchange Technique (IET),[2] based on measuring the accumulated analyte in a resin column; (3) The Resin Titration (RT) technique,[3] based on the competitive complexation of the metal ions to resin beads (*e.g.* Chelex) and total elemental concentration determination; (4) The Donnan Membrane Technique (DMT),[4] which exploits a cationic exchange membrane to obtain a simpler speciation system in the acceptor compartment; (5) The Polymer Inclusion Membrane devices (PIM),[5] based on the chemically-driven accumulation process in an acceptor cell through a membrane with a carrier; *etc.*

The dynamic techniques typically yield a flux. Among the dynamic techniques, one can mention: (1) Diffusion Gradients in Thin-films (DGT),[6] where the flux is usually reported in terms of an operationally defined concentration (c_{DGT}); (2) Voltammetric techniques,[7] such as Differential Pulse Polarography (DPP), Reverse Pulse Polarography,[8] Anodic Stripping Voltammetry (ASV), Cathodic Stripping Voltammetry[9] or Scanned Stripping ChronoPotentiometry (SSCP),[10,11] based on the impact of the potential on the intensity current; *etc.* One challenge in the dynamic techniques is to go beyond operationally-defined parameters.[12,13] The measurements in the dynamic techniques cannot be described just in terms of equilibrium concentrations, but one can interpret the results in terms of sound physicochemical parameters (such as diffusion coefficients, rate constants, equilibrium constants, *etc.*), so that, from the measurements, one can (in principle) retrieve (bulk) equilibrium concentrations of real species.[14–16] A typical difficulty in this endeavor is the lack of reliable parameters in complex systems (for instance, rate constants or diffusion coefficients of unknown or poorly characterized ligands in an environmental sample).

There are also some techniques, such as Permeation Liquid Membrane (PLM),[17] that can act either in an equilibrium or in a dynamic regime, depending on some controlled parameter (*e.g.* the hydrodynamic conditions).

It is clear that no technique can answer all the diverse needs in speciation studies. Sometimes, there is the need of *in situ* deployment in a river, of a great precision, of a fast response, of high selectivity, or of accessing to some special parameter, *etc.* Not all techniques are valid for all elements, and usually two techniques report different information, even if applied to the same element and sample. So, the different techniques are tools providing complementary information, and each one with its own advantages, limitations, and drawbacks.[13,18,19]

In this scenario, AGNES (Absence of Gradients and Nernstian Equilibrium Stripping) stemmed from the voltammetric family, but with a key shift towards the equilibrium paradigm. AGNES was developed with the aim of providing (bulk) equilibrium free metal ion concentrations, even at the range of picomol per liter, with a sound and simple interpretation of the analytical signals. In the necessary trade-off between characteristics, AGNES privileges robustness of interpretation at the expense of time (in designing and running experiments). An example of the used equipment in AGNES is shown in Figure 1. After the initial presentation of AGNES,[20,21] a few reviews have been published,[9,22–24] as well as a very detailed tutorial guide with specific protocols for a hands-on implementation of AGNES.[25] In line with the goal of this handbook, this chapter will start with the basics of the technique and will follow with some applications in a variety of natural and synthetic systems.

IV. Fundamentals of AGNES Technique

A. Principles of AGNES

AGNES is a stripping technique consisting in two conceptual stages: (1) preconcentration of an amalgamating element M (Zn, Cd, Pb, In, Sb, Sn,

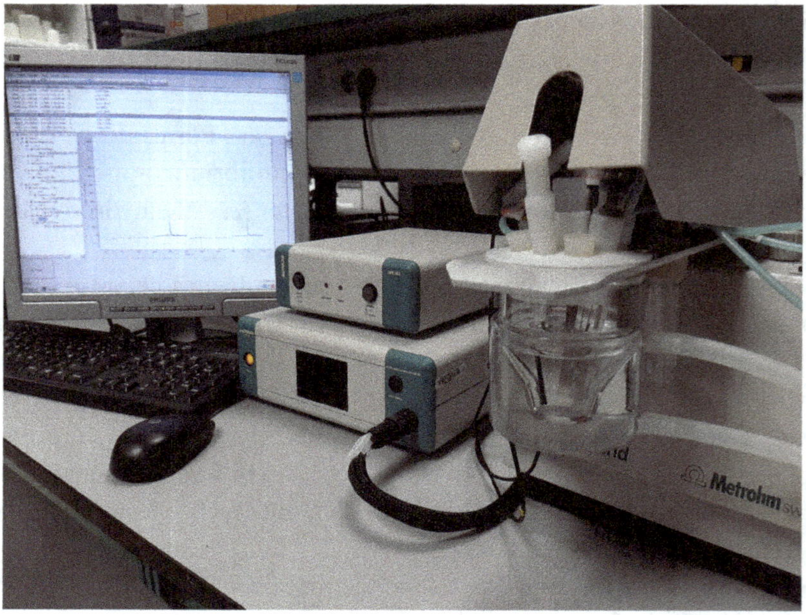

Figure 1. Typical equipment used for AGNES experiments with the polarographic stand and the potentiostat.

Ga, Tl) along a deposition stage and (2) the stripping of the accumulated analyte. The key difference with many other voltammetric techniques is that each stage has to fulfil a very specific goal as it is outlined in the next sub-sections below. One can combine any of the variants for the first stage (described in Section IV.B) with any of the variants for the second stage (described in Section IV.C).

1. *First stage: reaching equilibrium with a known gain*

The goal of the first stage is to accumulate the reduced form of the analyte (typically a metal), M^0, through the application of a reducing potential, until M^0 is in equilibrium with the free metal M^{n+} in the solution. This involves waiting as much as necessary until there is absence of gradients in the concentration profiles of all relevant species. Thus, by the end of the first stage, the concentration of M^{n+} in solution is homogeneously constant from the surface of the electrode up to any distance in the bulk (see

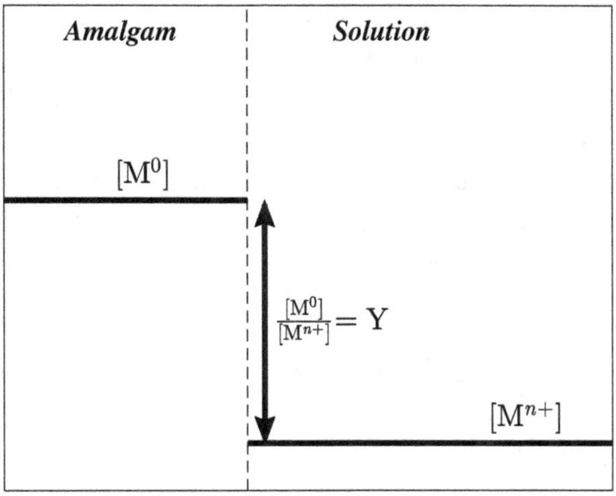

Figure 2. Aimed concentration profiles by the end of the first stage of AGNES. The analyte has been accumulated inside the electrode (left hand side), while the concentration is also homogeneous in the solution (right hand side). The ratio Y is the gain or preconcentration factor. Adapted from Ref. 21.

horizontal line on the right hand side of Figure 2) and the concentration of M^0, $[M^0]$, is constant anywhere in the mercury electrode (see horizontal line on the left hand side of Figure 2). The number of moles accumulated in the (small) electrode is negligible in front of the total number of moles in solution, so that the equilibrium concentration in solution, $[M^{n+}]$, is the same as the one before the application of AGNES. These aimed homogeneous concentrations imply that there is no flux of M across the electrode surface by the end of the first stage. So, concomitantly, AGNES requires that, then, there is also Nernstian equilibrium (of the couple M^{n+}/M^0) at the electrode surface. In Section IV.B, two strategies (1P and 2P) to practically reach this equilibrium will be presented, but both eventually apply the deposition potential E_1 (where subscript 1 indicates first stage).

Due to Nernst law, the gain (or preconcentration factor) Y, achieved by the end of the first stage and associated to E_1, can be expressed as:

$$Y \equiv \frac{\left[M^0\right]}{\left[M^{n+}\right]} = \exp\left[-\frac{nF}{RT}\left(E_1 - E^{0'}\right)\right] \tag{1}$$

where n is the number of exchanged electrons; F, the Faraday constant; R, the gas constant; T, the temperature and $E^{0'}$ is the formal standard redox potential of the couple. The list of symbols is gathered at the beginning of this chapter. To avoid the explicit knowledge of $E^{0'}$, the gain can be computed, for reversible couples (*e.g.* Zn, Cd, Pb in simple matrices), from an ancillary DPP experiment with a sufficiently large electrode (so that the equations for planar diffusion apply):

$$Y = \sqrt{\frac{D_{\mathrm{M}^{n+}}}{D_{\mathrm{M}^0}}} \exp\left[-\frac{nF}{RT}\left(E_1 - E_{\mathrm{peak}} - \frac{\Delta E}{2}\right)\right] \tag{2}$$

where $D_{\mathrm{M}^{n+}}$ is the diffusion coefficient for the free metal ion in solution, D_{M^0} is the diffusion coefficient for the reduced metal inside the amalgam, E_{peak} is the peak potential in the DPP polarogram and ΔE is the modulation amplitude (or pulse height).

To compute the gain for not fully reversible redox couples (*e.g.* In^0/In^{3+}), see Sections IV.E.3 and IX.

The concept of gain is crucial in AGNES. For each deposition potential E_1, there is one Y reached at equilibrium, as Eq. 1 indicates. Changing E_1 means changing the desired Y and *vice versa*. Moreover, deposition times and limits of detection depend on Y, so its determination (even if it was approximate) is very useful.

2. *Second stage: quantifying the accumulation*

The aim of the second stage is to quantify the amount of M^0 inside the amalgam. This can be done either by measuring a current under diffusion-limited conditions (for re-oxidation) or by measuring the total charge in the (complete depletion) stripping of accumulated M^0 (see variants of second stage in Section IV.B).

The concentration in the amalgam can be quantified, for instance, by applying a reoxidation potential under diffusion-limited conditions and measuring the current at a given fixed time t_2. When the faradaic current under diffusion-limited conditions I is obtained, (see Supporting Information, SI, in Ref. 21)

$$I = \eta\left[\mathrm{M}^0\right] \tag{3}$$

where η is a proportionality factor that can be found in a calibration. So, by combining Eqs. 1 and 3, the fundamental relationship for AGNES results in:

$$I = \eta Y \left[M^{n+} \right] \tag{4}$$

Analogously, for variants of the second stage that finally lead to the total faradaic charge Q, Faraday's law for an electrode of volume V_{Hg} implies:

$$Q = nFV_{Hg} \left[M^0 \right] = \eta_Q \left[M^0 \right] \tag{5}$$

and, by combining with Eq. 1, Eqs. 6 and 7 are obtained.

$$Q = \eta_Q Y \left[M^{n+} \right] \tag{6}$$

$$\eta_Q = nFV_{Hg} \tag{7}$$

The proportionality factors η or η_Q are usually found from an ancillary calibration where I or Q is plotted against known values of $[M^{n+}]$, *e.g.* computed with a speciation program, such as Visual MINTEQ[26] (VM), or *via* any other theoretical or experimental means (see Figure 3).

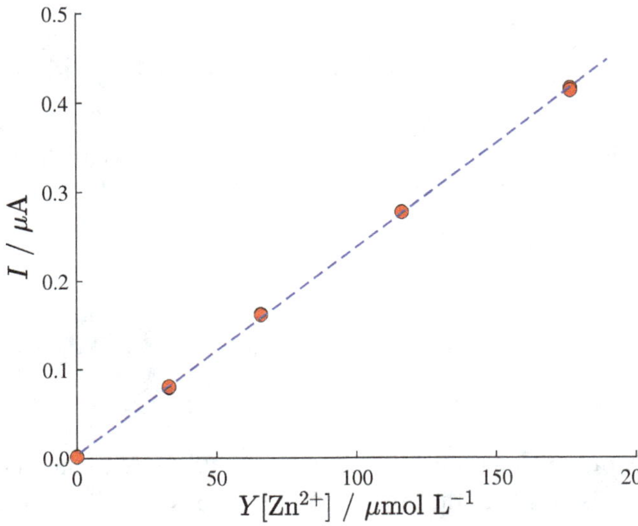

Figure 3. Calibration for Zn^{2+} in $[KNO_3] = 0.011$ mol L^{-1}, pH = 4.5 with the HMDE electrode, AGNES-1P, $Y = 17$, $t_1 = 200$ s, $t_w = 50$ s, and $t_2 = 0.2$ s. Determined η value 0.0023 A mol^{-1} L. Unpublished results.

The knowledge of Y needs not to be strictly accurate, because of the calibration procedure. Indeed, if the real Y value is assumed to be affected by a certain (unknown) multiplication factor, this same factor will appear both in the calibration and in the measurement, thus cancelling out. This is true even if one calibrates at one gain and measures using another gain (which are related *via* the Nernst equation).

The calibration is performed at the same temperature and ionic strength as those of the sample and covering the same range of $[M^0]$, so that gains different from those used in the calibration can be applied for the sample measurement.

The gains depend on the ionic strength because the Nernst equation (typically relating activities), when written in terms of concentrations as in Eq. 1, includes $E^{0'}$ which depends on the ionic strength.

If needed (*e.g.* if the sample's ionic strength is too low), a gain determined with DPP can be transferred from one ionic strength (μ_1) to another ionic strength (μ_2) with a correction involving the activity coefficients (γ) for M^{n+} in both media according to Eq. 8 (see also end of Section VII, on the application of AGNES to natural samples).[13,24,27,28]

$$Y^{\mu 2} = \frac{\gamma_{M^{n+}}^{\mu 2}}{\gamma_{M^{n+}}^{\mu 1}} Y^{\mu 1} \tag{8}$$

It is worth highlighting that AGNES is not impacted by the typical problem (in many voltammetric techniques) of a large ohmic drop at low ionic strengths, because when equilibrium is approached (by the end of the first stage), the current is negligible (just I_{Ox}) and leads to practically the same, very small, ohmic drop in calibration and measurement.[27]

B. Variants for the First Stage. Trajectory

1. *One pulse: AGNES-1P*

In the following explanations, it is assumed that before the application of AGNES there is no M^0 in the amalgam, but there is a (bulk) concentration $[M^{n+}]$ in the solution in contact with the electrode. The simplest potential program (*i.e.* input script written with a potentiostat control software such

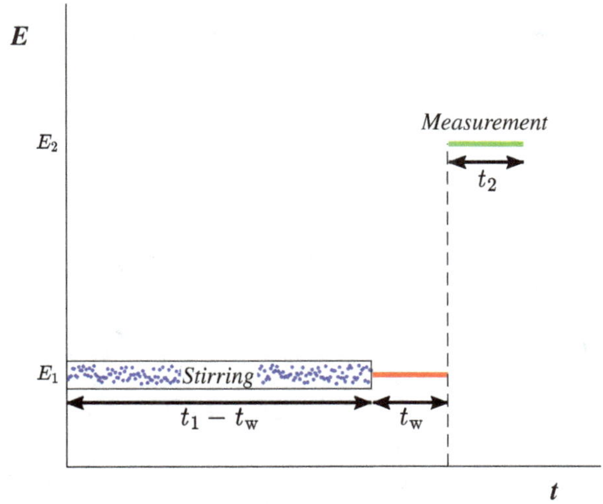

Figure 4. Schematic representation of the input potential program with variant AGNES-1P for the first stage and AGNES-I or AGNES-Q for the second one. The thick blue line indicates stirring. E_1 is the potential linked to the desired gain Y. Adapted from Ref. 20.

as GPES -Eco Chemie, The Netherlands or NOVA -Metrohm Autolab, The Netherlands) for the first stage consists in the steady application of the deposition potential E_1 for a sufficiently long time t_1 (see Figure 4). The potential E_1 is linked to a desired gain Y (see Eq. 1) and the system starts to evolve towards equilibrium by "pumping up" M across the electrode surface:

$$\mathrm{M}^{n+} + ne^{-} \rightarrow \mathrm{M}^{0} \tag{9}$$

Figure 5 simulates the evolution of the concentration profiles along the first stage,[21] as a result of the transfer of M from the solution (as M^{n+}) towards the amalgam (as M^0), while keeping bulk conditions constant at a fixed distance of the electrode surface (due to stirring). The profile of M^0 after 50 ms of application of the potential pulse (see darker blue dashed curve on the right of the electrode region $r < r_0$) indicates that this time has not been enough for a perceivable amount of M^0 to reach the center of the drop, while the gradient at the electrode interface is large,

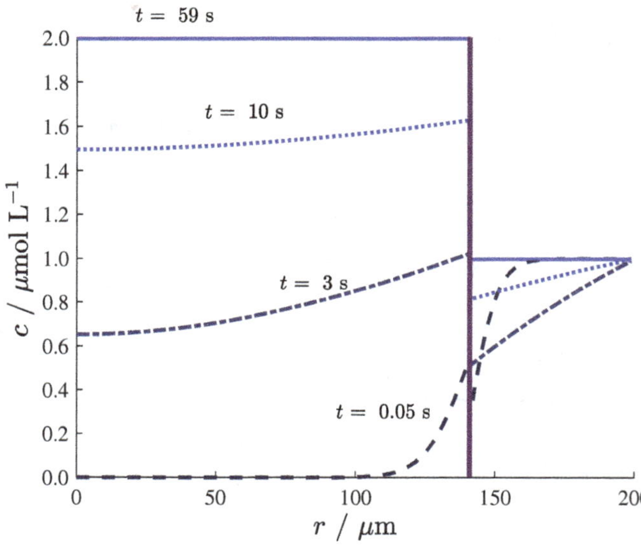

Figure 5. Evolution of the concentration profiles as simulated for a spherical electrode of radius r_0 = 141 μm (simulating drop 1 in HMDE). Y = 2, fully reversible redox couple, $D_{M^0} = 1.6 \times 10^{-9}$ m^2 s^{-1}, $D_{M^{n+}} = 7.3 \times 10^{-10}$ m^2 s^{-1}, bulk free metal concentration 1 μmol L^{-1} restored at r = 200 μm. Distances are measured with respect to the center of the mercury drop. The vertical purple line indicates the electrode interface. Unpublished data.

indicating a large flux and a large faradaic current, as follows from first Fick's law:

$$I = nFAD_{M^0}\,\text{grad}\left[M^0\right] \tag{10}$$

The profile of M^0 at 3 s (dotted-dashed line) has already reached the center and starts a general accumulation, while now the gradient at the interface is less than that at 50 ms. Other profiles (in progressively lighter blue colors in Figure 5) follow the same trend, so that at 59 s one can accept that the desired absence of gradients in the concentration profiles is fulfilled, not only for M^0, but also for M^{n+} in the solution.

The flux of M is larger at the initial times (*i.e.* start of the deposition) and keeps decreasing with elapsing time, until it becomes practically null, when equilibrium is approached. So, according to Eq. 10, the measured intensity current along the deposition stage will keep decreasing in absolute value, as seen experimentally (see Figure 6). Since an enhanced mass

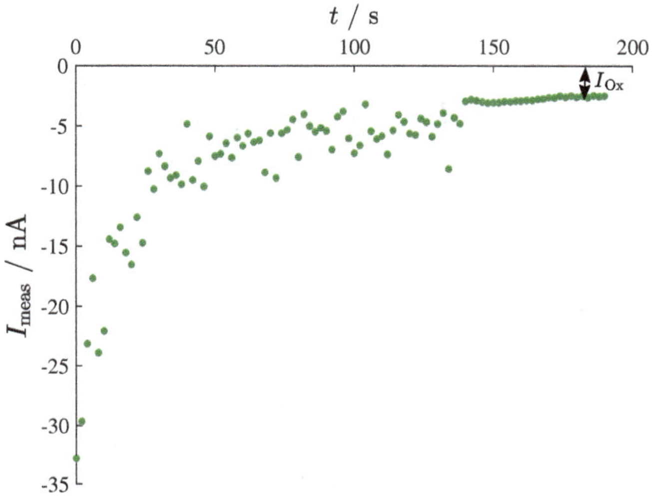

Figure 6. Intensity measured along the first stage of AGNES-1P. No stirring for the last $t_w = 50$ s allows also to estimate I_{Ox}. $c_{T,Zn} = 10.6$ μmol L^{-1} in 0.011 mol L^{-1} KNO_3 pH = 4.5 with the HMDE electrode, AGNES 1P, $Y = 17$, $t_1 = 140$ s, $t_w = 50$ s and $t_2 = 0.2$ s. Unpublished data.

transport across the solution decreases the needed deposition time, during most of deposition stage one applies some stirring (see thicker blue bar with scattered inside points in Figure 4), which introduces a relevant noise in the measured current (see Figure 6). In some cases (high analyte concentrations), the approach to equilibrium can be seen as an eventual stabilization of the noisy deposition currents towards an approximately constant value (*e.g.* around –5 nA in Figure 6). Stirring is stopped a few seconds (t_w) before the end of the first stage (see thin red line in Figure 4) to increase the reproducibility associated to measurement of the second stage. The eventual recorded current, labelled as the oxidants current I_{Ox}, see Figure 6, is usually associated to the unavoidable traces of oxygen in the voltammetric cell (despite purging with nitrogen or argon).[29]

The attainment of the equilibrium sought in the first stage can only be confirmed by performing different complete AGNES runs with different deposition times. The collection of the analytical responses (I or Q) from AGNES experiments (each one with its own first and second stage) with different deposition times, for a fixed gain Y, is called a trajectory. Blue

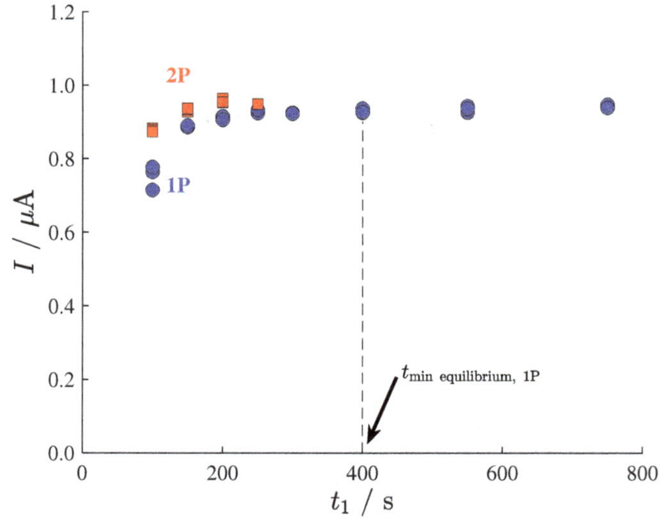

Figure 7. Simple trajectories of the faradaic intensity current at $Y = 50$ with AGNES-1P (blue bullets) or AGNES-2P (red squares). The vertical dashed line indicates an approximate minimum deposition time to achieve equilibrium with AGNES-1P in this system. Experimental settings: $Y = 50$ (1P), $Y_2 = 10^{-8}$, $Y_{1,a} = 10^{10}$, $Y = 50$ (2P), $t_w = 50$ s, $t_2 = 0.2$ s, $c_{T,Zn} = 10.1\ \mu$mol L^{-1}. Adapted from Ref. 20.

bullets in Figure 7 show a trajectory for $Y = 50$, where stabilization (associated to equilibrium) is reached only at longer deposition times (say from 400 s onwards). Shorter times (*e.g.* $t_1 - t_w = 100$ s) have not been enough to reach equilibrium and correspond to a situation called "undershoot".

Further details on the question "How long should one wait?" will be addressed in Section IV.E.

2. *Two pulses: AGNES-2P*

In the 2 pulses strategy, to be considered when the necessary deposition times with AGNES-1P are excessively long, the deposition stage is split into two potential sub-stages.[20] In the first sub-stage, lasting for a time $t_{1,a}$, a potential $E_{1,a}$ corresponding to diffusion-limited conditions for reduction is applied (see Figure 8). In the second sub-stage, lasting $t_{1,b}$, one applies the deposition potential E_1 (which could also be called $E_{1,b}$) corresponding to the desired gain Y. The first sub-stage accumulates M^0 at

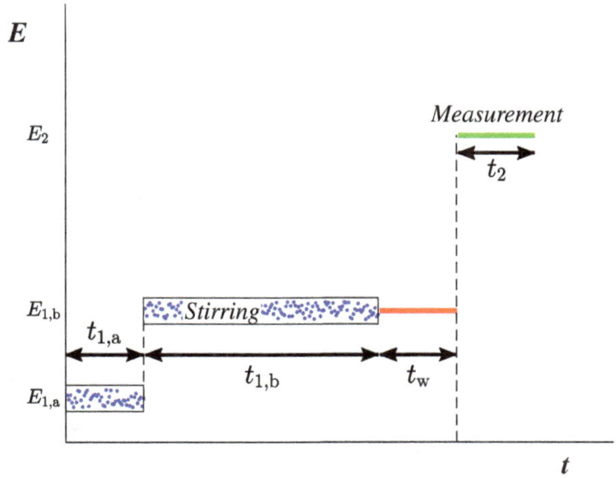

Figure 8. Input potential program with variant 2P for the first stage and AGNES-I or AGNES-Q for the second one. $E_{1,b}$ is the potential linked to the desired gain Y. In AGNES-Q the total stripped charge along t_2 is computed by integrating the current intensity. Adapted from Ref. 20.

the maximum possible speed. If $t_{1,a}$ is judiciously chosen, the deposition time can be reduced typically by a factor of 10. For non-optimal $t_{1,a}$-values, one can encounter two situations:

— Undershoot. If $t_{1,a}$ is too short for the required gain, there is, by the end of the first sub-stage, still more M^0 missing inside the electrode for equilibrium. One needs to have a sufficiently long second sub-stage (during $t_{1,b}$, applying E_1) to reach equilibrium. For high concentrations, the undershoot can even be seen along the deposition stage (see green diamonds in Figure 9): when the second stage starts ($t > 15$ s), the current still keeps reducing its absolute value as equilibrium is approached.

A much general, and safer, diagnose that a certain $t_{1,a}$ leads to undershoot can be derived from a trajectory with fixed $t_{1,a}$ (and Y): the analytical response (Q or I) keeps increasing for increasing $t_{1,b}$, as illustrated in the series with red squares in Figure 7 or in the series with green bullets or blue diamond markers (notice that $t_{1,a} = 0$ is equivalent to AGNES-1P) in Figure 10.

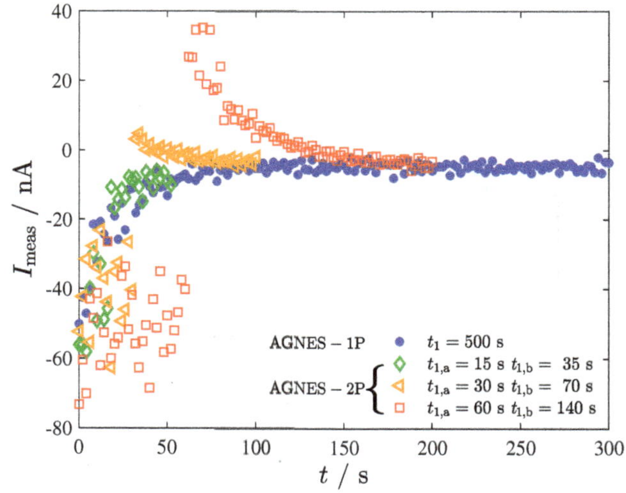

Figure 9. Overshoot and undershoot as seen in the measured currents along the deposition stage for a sufficiently concentrated solution. $c_{T,Zn} = 10.1$ μmol L^{-1}, $Y = 50$, $Y_{1,a} = 10^{10}$, pH = 5.0, HMDE. Unpublished data.

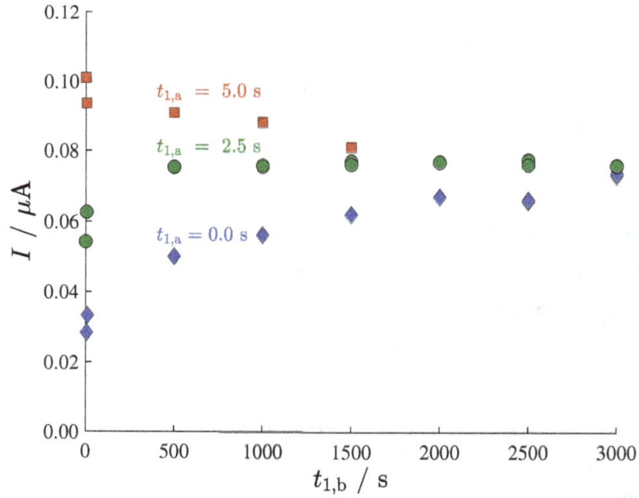

Figure 10. Overshoot and undershoot easily diagnosed from trajectories of the faradaic current. The trajectory labelled $t_{1,a} = 0$, blue diamond markers, indicates the strategy AGNES-1P for the first stage (naturally linked to "undershoot"). Initial points in the trajectory labelled $t_{1,a} = 2.5$ s, green bullets markers, are "undershoots," since $t_{1,a}$ was too short and further accumulation is needed. Initial points in the trajectory labelled $t_{1,a} = 5$ s, red square markers, are "overshoots" in the AGNES-2P variant for the first stage. The sample was Raimat Abadia (Catalonia, Spain) red wine. Adapted from Figure 3 in Ref. 32.

— Overshoot. If $t_{1,a}$ is too long for the prescribed Y, one has accumulated too much M^0 inside the electrode and one needs suitable $t_{1,b}$ to relax back to the equilibrium $[M^0]$. For high concentrations, this can be seen as a jump in the current at the transition between sub-stages and a decay of the deposition currents from currents more positive than I_{Ox} (see orange triangles and red squares in Figure 9 for $t > 30$ and 60 s, respectively). Notice that in the first sub-stage (*e.g.* $t < 60$ s in the red square series), the deposition current is roughly constant (due to the approximate steady-state accumulation affected by stirring). Again, the safe diagnose comes from analytical signals being progressively lower for longer $t_{1,b}$ (for fixed given $t_{1,a}$ and gain), as illustrated in Figure 10, red squares.

As a general rule, it is recommended to satisfy the relation 11:

$$t_{1,b} = 3t_{1,a} \tag{11}$$

to have a suitable relaxation of possible undershoots or overshoots.[20] However, in some cases, shorter $t_{1,b}$ values can be useful,[30,31] such as:

$$t_{1,b} = t_{1,a} \tag{12}$$

since modeling (with some typical assumptions) indicates that the maximum overestimation of the concentration would be just around 13.5%.

Workflows to adjust deposition times and gains have been published,[28,30] but many other strategies are possible.

C. Variants for the Second Stage

1. *AGNES-I*

In this variant,[21] the analytical signal is the faradaic stripping current I_{meas} measured at a fixed time (t_2 for measurement is usually 0.2 s with the HMDE, to minimize the capacitive current impact) of a potential step at a stripping potential E_2 corresponding to diffusion-limited conditions for the reoxidation of M^0 back to M^{n+} (see green lines in Figures 4 or 8).

The contribution of oxidants can be discounted from the "raw" read current, by subtracting from it the eventual residual or "limit" current, I_{lim}, at the end of the second stage (usually averaging the currents from 48 to 50 s, in the stripping stage).

This obtained faradaic current (at t_2) is very accurate, since it is not disturbed by interferences due to adsorption on the electrode, hydrodynamic conditions, or concomitant homogeneous reactions in the solution whenever reoxidation in diffusion-limited conditions prevails.[21,24] Eq. 4 applies with I being the faradaic current. As the measured current I_{meas} (at the equilibrium plateau) has other components such as the oxygen current or a capacitive current, a blank needs to be subtracted from the total measured current. The simplest blank is the one performed in a synthetic solution without the analyte M (*i.e.* just the background electrolyte). Other strategies for blank correction are commented in Ref. 32 and in Section VII.

The variant AGNES-I is very convenient for elements such as Zn, whose standard redox potential is far away from those of other elements (such as Cd).

2. *AGNES-Q*

In AGNES-Q, the analytical signal is the charge stripped until complete depletion of M^0 in the amalgam, along a potential step not necessarily corresponding to diffusion-limited conditions for reoxidation (see the second stage in Figures 4 or 8). So, the constant potential step at E_2 has to be applied for a sufficiently long time (*e.g.* $t_2 = 50$ s with the HMDE) in order to achieve complete depletion[33] and quantify all accumulated M^0, (see Figures 4 or 8). The charge is usually computed by integration of the current over time. A suitable blank, such as the measurement in a synthetic solution, must be subtracted to the total charge to obtain the faradaic charge which is the one fulfilling the fundamental Eq. 6.

This variant is suitable for elements such as In, whose standard redox potential is just slightly more negative than that of Pb, and applying the variant AGNES-I could result in the interfering reoxidation of accumulated Pb.[34] Also in systems (*e.g.* with low ionic strength), where the stripping current along a pulse is anomalous.[35,36]

3. *AGNELSV*

In AGNELSV (Absence of Gradients and Nernstian Equilibrium with Linear Stripping Voltammetry) the faradaic charge Q is obtained from the area below the peak of an I *vs.* E curve (and above the baseline) appearing when the stripping potential (E) is linearly scanned from E_1 to a sufficiently more positive E_2. The scan rate must be sufficiently slow as to have complete depletion of M^0. No blank is required.[24] The fundamental Eq. 6 can then be applied.

This variant has a similar prescription to that of AGNES-Q, with the added advantage of not requiring a blank.

4. *AGNES-SCP*

In this variant of the second stage, a (small) fixed stripping current (I_s) is imposed, and the potential evolution is recorded, until a sudden variation in the potential, in principle,[92] indicates the fulfilment of complete depletion of M^0. *Via* a change of representation of the data to dt/dE *vs.* E, the so-called transition time (τ) can be measured from the area below the peak and above the baseline.[37] The faradaic charge can be accurately computed[38] from Eq. 13, where I_{Ox} is the oxidants current (measured at the end of the deposition stage, see Figure 6). In some instances, I_{Ox} can be neglected in front of I_s.

$$Q = \left(I_s - I_{Ox}\right)\tau \tag{13}$$

AGNES-SCP is accurate and the most suitable to recognize and eliminate interferences. For instance, in Figure 11 one can clearly distinguish the peak of Cd (the analyte) from the peak of Pb (in this case, the interferent). When stabilization is reached and further deposition times are probed, the peak of Cd stays fixed, while the one of Pb increases.

D. Working Electrodes

1. *Mercury electrodes*

The typical implementation of AGNES has been conducted with mercury electrodes. A general rule is that the smaller the size of the electrode, the faster a certain gain can be reached, as easily demonstrated.[21]

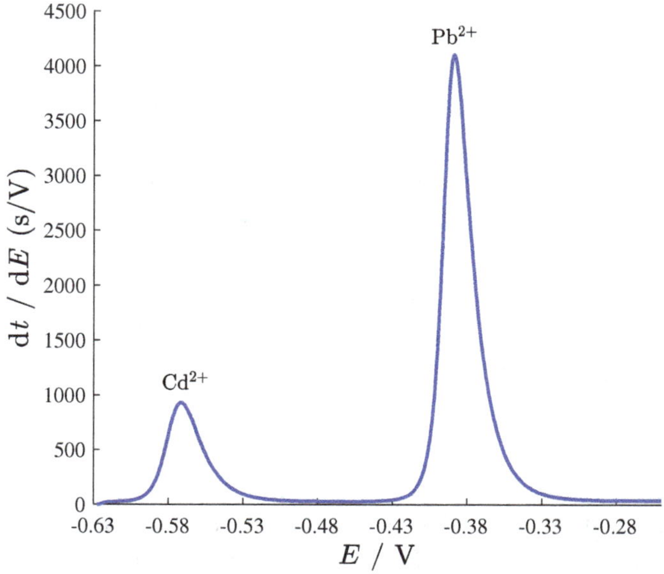

Figure 11. AGNES-SCP processed data showing the resolution capacity of the technique in the determination of $[Cd^{2+}]$ with the presence of the interferent Pb^{2+}.[38] $c_{T,Pb} = c_{T,Cd} = 50$ nmol L^{-1}, $Y = 500$, HMDE, $t_{1,a} = 350$ s, $t_{1,b} = 1050$ s, $t_w = 50$ s, $I_s = 10^{-9}$ A. Unpublished data.

(a) HMDE (Hanging Mercury Drop Electrode)

The main advantage of HMDE is that a new drop is generated for each AGNES run, which means getting rid of possible accumulated adsorption interferences on the electrode surface. Even though, by design, AGNES is quite unaffected by adsorption,[24] extreme cases can eventually block the electrode, which, in practice, can be seen as a kind of electrodic irreversibility. For instance, the renewable feature of the HMDE was key[24] to tackle the recalcitrant adsorption of a nanoparticle dispersant (see Section VIII).

On the other hand, the smallest drop delivered by a standard polarographic stand (such as Metrohm, with a radius around 141 μm) is larger than those of the mercury electrodes discussed next, and so a drawback of the HMDE is the typically longer required deposition times for a fixed desired gain.

(b) Iridium plated microElectrode (Ir-μE)
A 5 μm-radius iridium wire can be coaxially embedded into a glass cylinder to expose a disc Ir-electrode coplanar to the glass cylindrical lower surface. On top of the Ir disc, a drop of Hg can be plated. Although the ratio of capacitive over faradaic current increases (in comparison with the HMDE), lower free concentrations can be accessed for a fixed time.[39] One potentiality of such microelectrodes is the possibility of working with smaller volumes of sample and, ideally, being introduced in living organisms such as a plant xylem. One limitation of this microelectrode is the need of carefully polishing the co-planar surfaces (to obtain reproducible results) and the inherent fragility of the glass cylinder.

(c) Screen-Printed Electrode (SPE)
SPEs are prepared by painting a conductive (black carbon) ink on top of a plastic substrate.[35] Then, mercury can be plated from an Hg(II) solution, as with the Ir-μE. No polishing is needed. One advantage of the SPE is its disposable nature and its consideration as more environmentally acceptable (given the very small amount of Hg involved). SPE can be used for on-site measurements (see Section VII). One minor drawback is the relatively large capacitive current, rendering the variant AGNES-I not suitable with this kind of electrode.

(d) Thin Mercury Film Rotating Disc Electrode (TMF-RDE).
A glassy carbon disc electrode is coaxially embedded in a rotating shaft. After polishing and cleaning, mercury can be plated on the disc electrode. Not all the carbon surface is electroactive, so that a series of Hg^0 droplets randomly grow on it, rendering the term "film" just indicative, not descriptive. The combination of very small radii (of the order of a few nm) and the lowered mass-transport limitation (due to the fast rotation of the ensemble) allows the TMF-RDE to be the best performant (in comparison with other electrodes) to reach extremely low free metal ion concentrations.[13,40,41] One drawback of the TMF-RDE is the danger of some droplets falling down, modifying the effective volume of Hg.[42] The coating of the mercury film with a Nafion layer prevents protein adsorption on it, thus enabling, *e.g.*, the measurement of [Zn^{2+}] in the presence of bovine serum albumin (BSA) and in a sample of fetal bovine serum (FBS).[25]

2. *Non-mercury electrodes*

AGNES was originally designed just for mercury electrodes while targeting amalgamating elements and most AGNES publications rely on such electrodes. Thus, for the sake of simplicity, the condition of equilibrium for the first stage in AGNES has been detailed, in Section IV.A.1 above, with Hg electrode and an amalgamating element. However, the equilibrium condition can also be fulfilled on the surface of a non-Hg electrode if a monolayer can be partially occupied. Details on the theoretical background can be found in Ref. 43.

(a) Bismuth Film Electrode (BiFE)
Ex situ plated Bi film electrodes (BiFE) achieved the quantification of Pb^{2+} after checking the attainment of AGNES conditions *via* trajectories and the proportionality between gain and the stripped charge (from AGNES-SCP).[43] A threshold in the maximum operative charge (probably linked to the saturation of the monolayer with "deposited" Pb^0) was found. Speciation results were checked with the ligands polystyrenesulfonate and iminodiacetate.

(b) Gold electrode
AGNES was applied with the Vibrating Gold Microwire Electrode (VGME) in the determination of Cu^{2+}.[44] Copper speciation was checked with malonate and iminodiacetate ligands.

No further comments on these non-Hg electrodes are done in the rest of this chapter.

E. Deposition Times

1. *Rules of time in solutions with just the free analyte*

Guidelines on the order of magnitude of the expected deposition time are very useful to plan the work, to choose the electrode type, *etc*. Rules have been derived for the case of just the free ions in solution (*i.e.* without complexes). These "rules of thumb" can, later on, be extended by taking into account the contribution of the complexes (see Section IV.E.2).

Obviously, the larger the gain, the longer one must wait, as more accumulation is prescribed. So, rules will imply the gain Y as the essential factor.

As commented above, the larger the size of the electrode, while keeping all other factors equal, the longer the required time to reach a certain gain. This was theoretically established [21] and experimentally observed since the first application of AGNES. So, rules depend also on the electrode size.

The derivation of the rule can be done from observation of the approximate deposition time required for an acceptable resulting analytical response (*i.e.* practically that of the equilibrium plateau) with a given Y. A more systematic procedure consists in building up trajectories with different gains (in the same solution) and re-plotting them in terms of Q/Y *vs.* (deposition time)/Y. This normalization leads to a merging (or collapse) of all trajectories in a master trajectory from which the extraction of the rule is straightforward,[34] as illustrated in Figure 12.

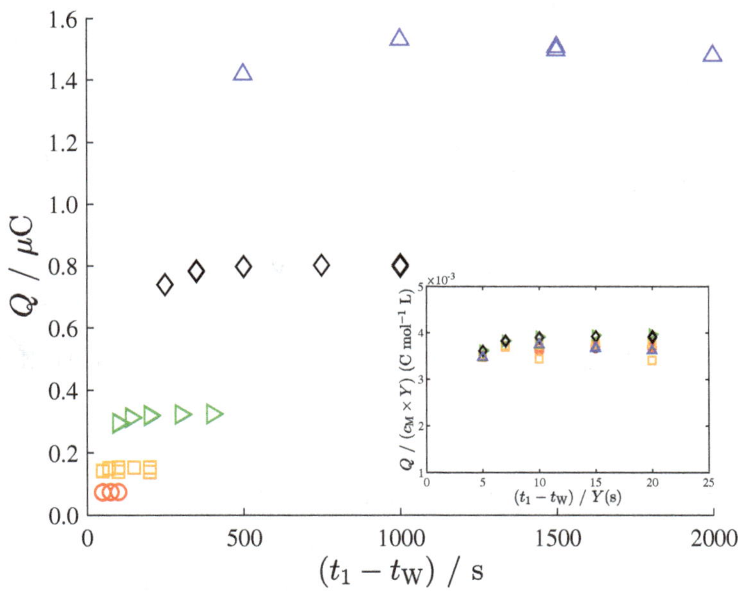

Figure 12. Original trajectories and collapse of their normalized version (inset) for indium with $c_{T,In} = 5.00$ μmol L^{-1} at pH = 3, with HMDE. Gains: 5 (red circle markers), 10 (orange square markers), 20 (green triangle markers), 50 (black diamond markers) and 100 (blue triangle markers). From the normalized plot, the rule of time given by Eq. 19 was derived. Adapted from Ref. 34.

For Pb/Zn/Cd (*i.e.* reversible couples when there is just supporting electrolyte besides the target ion), with standard stirring (level "6" in Metrohm stand, corresponding to around 3000 min^{-1}) in the HMDE, the deposition time required with one pulse in the deposition stage (*i.e.* AGNES-1P variant, see Figure 4) is given by Eq. 14:

$$t_1 - t_{\mathrm{w}} = 7 \times Y \tag{14}$$

with the result (here and from now on, for these rules) expressed in seconds.[24,43]

In the case of Zn with RDE (at a rotation speed of 1000 min^{-1}), the rule for AGNES-1P deposition has been estimated around:

$$t_1 - t_{\mathrm{w}} = 0.008 \times Y \tag{15}$$

which indicates that, for a common gain, RDE is *ca.*1000 times faster than HMDE.[13] However, in practice, the leverage of the LOD will be less remarkable because the analytical signal is also smaller due to the smaller Hg volume.

Rules for the optimization of $t_{1,\mathrm{a}}$ (*i.e.* the deposition time under diffusion limited conditions in AGNES-2P) can also be derived.[45] For instance, for Pb/Zn/Cd with the HMDE, Eq. 16 applies,

$$t_{1,\mathrm{a}} = \left(\frac{\delta r_0}{3 D_{\mathrm{M}}} \right) Y \approx Y \tag{16}$$

where δ is the diffusion layer thickness around the drop of the HMDE (of radius r_0).

2. *The impact of complexes. Helping factor*

Given a fixed M^{n+} concentration in the sample solution, the concomitant presence of complexes cannot interfere in the eventual retrieved analytical signal at equilibrium (*i.e.* the plateau in the trajectory). Indeed, the equilibrium value of $[M^0]$ depends only on $[M^{n+}]$, due to Nernst law (Eq. 1), regardless of other equilibria involving $[M^{n+}]$ (say to form one or more complexes).[11]

However, the time needed to reach equilibrium is affected by the presence of complexes, if they are totally or partially labile (*i.e.*, if they can dissociate in the diffusion layer close to the electrode, while the complex travels towards the surface of the electrode). Here the term "labile" is understood not as a thermodynamic concept (*i.e.*, weak), but as a kinetic one (*i.e.*, fast). Indeed, unless a given complex ML is fully inert (regarding the homogeneous reaction), during the first stage of AGNES, ML will be forced to dissociate due to the consumption of M^{n+} at the electrode surface and the ensuing generation of a gradient of $[M^{n+}]$ from the electrode up to the bulk of the solution. The larger the extent of ML dissociation at the electrode surface, the more labile is this complex. So, lability is related to the competing processes of reaction and diffusion. The concept of lability degree (ξ) has been introduced to quantify the contribution of complexes to the supply of M arriving at an (electro)active surface.[46–48] The lability degree of a complex ML is defined as the ratio between the actual contribution of ML to the flux and the maximum possible contribution to the flux if this complex ML was fully labile and can be computed as:

$$\xi = \frac{J - J_{\text{inert}}}{J_{\text{labile}} - J_{\text{inert}}} \qquad (17)$$

where is J the actual flux in the system, J_{inert} the flux if considering only the free metal (which is also the case of a totally inert complex) and J_{labile} is the flux that would be obtained if ML was fully labile. J can be measured directly, while J_{inert} and J_{labile} can be estimated from calibrations with no ligand L in solution.[13] The lability degree ranges from $\xi = 0$ for totally inert complexes to $\xi = 1$ for fully labile ones. If the deposition potential is very negative, it might even force ML to be reduced directly to M^0 (releasing the ligand L) in a phenomenon called "second wave" and, then, $\xi = 1$ (see Ref. 45).

The flux in a system with a certain complex increases by a "helping factor" $\left(1+\frac{D_{\text{ML}}[\text{ML}]^*}{D_{\text{M}^{n+}}[\text{M}]^*}\xi\right)$ with respect to the contribution of the free metal alone.[45,48] The superscript * has been added here to remind that these are bulk concentrations. Hence, the estimated deposition time, in any rule, is reduced by this helping factor. For instance, with HMDE, Eq. 14 for Zn/Cd/Pb rewrites as Eq. 18.

$$t_1 - t_\mathrm{w} = \frac{7Y}{1+\dfrac{D_\mathrm{ML}[\mathrm{ML}]^*}{D_{\mathrm{M}^{n+}}[\mathrm{M}]^*}\xi} \tag{18}$$

Usually, the lability degree is not known forehand, but one can suggest educated guesses for $[\mathrm{ML}]^*$, $[\mathrm{M}]^*$ and the ratio of diffusion coefficients. So, one can compute the time for two extreme cases: with no help at all (*i.e.*, just by using Eq. 14) and with the maximum possible help (using $\xi = 1$ in Eq. 18) and plan the work ahead with this information. There are theoretical analytical expressions to compute lability degrees from physicochemical parameters such as the dissociation rate constant(s) of the complex(es), the diffusion coefficient(s), the stability constant(s) and the diffusion layer thickness.[46,47,49–54]

3. *Irreversible systems*

AGNES requires Nernstian equilibrium between the free metal in solution and the reduced metal in the amalgam to be reached. So, AGNES would not work for elements whose dispersion in Hg is not a real amalgam, which is likely to be the case for Ni, Co, or Fe.

AGNES would require prohibitively long times for redox couples with an extremely slow electron transfer (*i.e.* irreversible redox couples) not having alternative pathways for the global conversion of M^{n+} into M^0. Indium might have been one candidate for such difficulty, but, for pH-values comprised between 3 and 7, there is a non-negligible fraction of hydroxy species (*e.g.* $InOH^{2+}$), which provides an alternative pathway *via* the fast conversion of In^{3+} to the labile hydroxy species and their fast reduction to In^0 (*e.g.* the couple $In^0/InOH^{2+}$ is reversible).[11,55] For pH values below 5.5, AGNES experiments did not show a relevant delay to reach equilibrium. Indeed, from the collapse seen in Figure 12 (or in Figure 4 of Ref. 34), Eq. 19 instead of Eq. 14:

$$t_1 - t_\mathrm{w} = 10Y \tag{19}$$

applies at pH 3. This change can be interpreted as follows: the timescale of the irreversibility (perhaps of the order of seconds) is practically negligible in comparison with the timescale of the attainment of absence of gradients in the concentration profiles (of the order of tens of minutes) which is strongly dependent on the values of the diffusion coefficients).[11,34]

However, a practical irreversibility can arise from redox couples (such as Zn^0/Zn^{2+}) that are quite reversible in media with simple ions (such as perchlorate, nitrate, chloride, *etc.*), but change their behavior in some complex matrices. This has been found in the determination of the Zn^{2+} concentration in wine[28,32] or in a nanoparticle dispersion[24] (see Section VIII). By applying suitable strategies, one can confirm that the two-fold equilibrium is finally reached: (i) stabilization of trajectories (if possible, checking the convergence of trajectories with over- and undershoot) and (ii) fulfilment of Nernst relationship (*i.e.,* checking more than one gain).

V. Validation of AGNES in Simple Solutions

The determination of free metal concentration with AGNES has been validated in different ways, starting with simple synthetic solutions. Speciation codes like Visual MINTEQ[26] have proved the validity of the technique for the systems Zn-NTA (Zn-nitrilotriacetic acid),[20,56] Zn-EDTA (Zn-ethylenediaminetetraacetic acid) and Zn-oxalate,[20,57] Cd-iodide, and Pb-xylenol orange;[24] the code SolEq[58] for the systems Pb-NTA and Pb-PDCA (Pb-pyridinedicarboxylic acid); [45] and MEDUSA[59] for the systems Pb-PDCA and Cd-PDCA.[60]

The free concentration with AGNES has also been compared to values measured with other electroanalytical techniques, like ISE (Ion Selective Electrodes) for the systems Pb-NTA and Pb-PDCA,[45] Cd-polyacrylic acid, Cd-iodide and Pb-xylenol orange,[24] or SSCP (Scanned Stripping ChronoPotentiometry) for the systems Pb-PDCA and Cd-PDCA.[60] Also non-electroanalytical techniques have proved the validity: DMT (Donnan Membrane Technique) for the system Zn-NTA[56] and RT (Resin Titration) for the systems Pb-NTA and Pb-PDCA.[45]

These expound validations have, thus, covered metals whose electron transfer process is not fully reversible (case of Zn^{2+} in some matrices), which undergo precipitation (Zn^{2+}+oxalate), which form either labile (Pb-PDCA) or practically inert (Pb-NTA) complexes, or whose

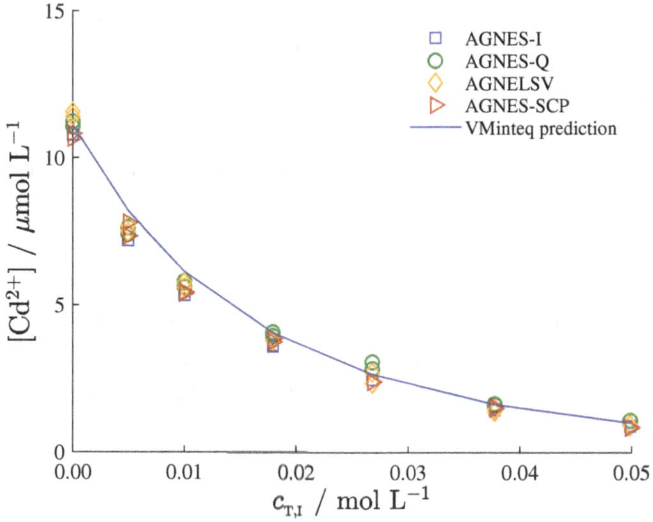

Figure 13. Variation of the free Cd^{2+} concentration along a titration with iodide at pH = 6 with different variants of AGNES (see internal legend). The initial total Cd concentration was 1.24×10^{-5} mol L^{-1}. Adapted from Ref. 24.

electroactive species adsorbs on the electrode (Cd-iodide, Cd-PAA, and Pb-xylenol orange).[24]

For the system Cd-iodide,[24] Figure 13 gathers the free metal concentration in a solution with increasing additions of iodide, measured with four different variants of AGNES (AGNES-I, AGNES-Q, AGNELSV, and AGNES-SCP, see Section IV.C). They all show a good concordance between them and with the predictions of VM.

Figure 14 shows the experimentally measured free cadmium concentration for the system Cd-polyacrylic acid (PAA)[24] at different pH values. Noteworthy, in a broad range of concentrations, the free Cd^{2+} concentration measured by AGNES or by a Cd-ISE are in very good agreement.

The speciation of Zn(II) in the presence of glutathione (GSH) was studied *via* titrations either adding the metal cation or the ligand. A comparison of the $[Zn^{2+}]$ measured by AGNES with the predictions resulting from published stability constants, allowed to support or discard models for the Zn binding to glutathione (GSH),[61] see Figure 15. Notice that, in this figure, the maximum free concentration of Zn is always below the

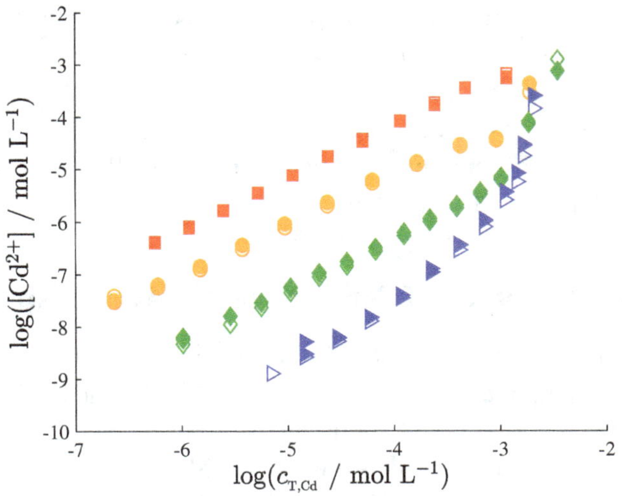

Figure 14. Variation of the free Cd^{2+} concentration in a solution containing polyacrylic acid at $c_{PAA} = 5 \times 10^{-3}$ mol L^{-1} and $[KNO_3] = 0.1$ mol L^{-1}. Red square markers for pH 4; orange circle markers for pH 5; green diamond markers for pH 6 and blue triangle markers for pH 7. Full markers correspond to AGNES and empty markers correspond to ISE measurements. Adapted from Ref. 24.

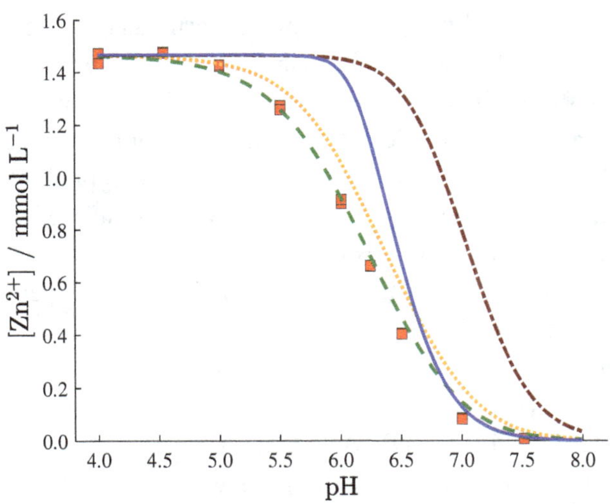

Figure 15. Discrimination between competing models for Zn-glutathione complexation. Red square markers: AGNES results. Models: green dashed line (Krezel *et al.*[93]), orange dotted line (Ferreti *et al.*[94]), blue continuous line (Diaz-Cruz I[95]) and dotted-dashed brown line (Diaz-Cruz II[95]). $c_{T,Zn} = 1.6 \times 10^{-3}$ mol L^{-1}, $c_{T,GSH} = 2.9 \times 10^{-3}$ mol L^{-1}, and KNO_3 0.1 mol L^{-1}. Adapted from Ref. 61.

total one $c_{T,Zn}$ due to the formation of ion pairs and complexes (*e.g.* with the nitrate anion of the supporting electrolyte).

Apart from all the commented validation procedures applied with HMDE, the system Pb-PDCA has also been checked with an Ir-μE,[39] the system Pb-IDA (Pb-iminodiaceticacid) with a TMF-RDE,[42] and the systems Zn-oxalate and Cd-oxalate with an SPE.[35]

VI. Application of AGNES in Humic Acid Dispersions

One of the main applications of AGNES has been the characterisation of metal binding to humic and fulvic acids. The free Cd^{2+} and Zn^{2+} concentrations measured by AGNES, along titrations of purified Aldrich humic acid with metal, at pH values ranging from 4 to 7, and the subsequent interpretation by means of the Non Ideal Competitive Adsorption (NICA) – Donnan (with a monomodal distribution) model, provided insights on the binding properties of both cations.[62]

The free Pb^{2+} concentration along titrations of purified Aldrich humic acid at a broader range of pH (4 to 9) was determined with relatively short deposition times for quite low concentrations (even sub-nanomolar) due to the lability and mobility of the complexes formed. The interpretation of the data through a bimodal NICA isotherm, see Figure 16, led to the Conditional Affinity Spectrum (CAS) at each pH and helped in the elucidation of the binding properties of these natural heterogeneous ligands.[31]

Similarly, Pb^{2+} binding to the three fractions (hydrophobic, transphilic and hydrophilic) extracted from the Dissolved Organic Matter (DOM) from the treated effluent of a wastewater treatment plant and from the Seine river was evaluated. The specific complexation properties of each fraction was determined.[30]

Cd^{2+} and Zn^{2+} binding to Laurentian Fulvic acid was also analyzed[63] with AGNES at pH 6, 7, and 8, in order to be able to describe further the binding properties with the NICA-Donnan model coupled to the speciation software ORCHESTRA[64] with the parameter estimation program PEST.[63]

The speciation of Cd^{2+}, Zn^{2+} and Pb^{2+} for six DOM samples (categorised into autochthonous and allochthonous sources, based on their absorbance and fluorescence properties) was evaluated with AGNES.[65]

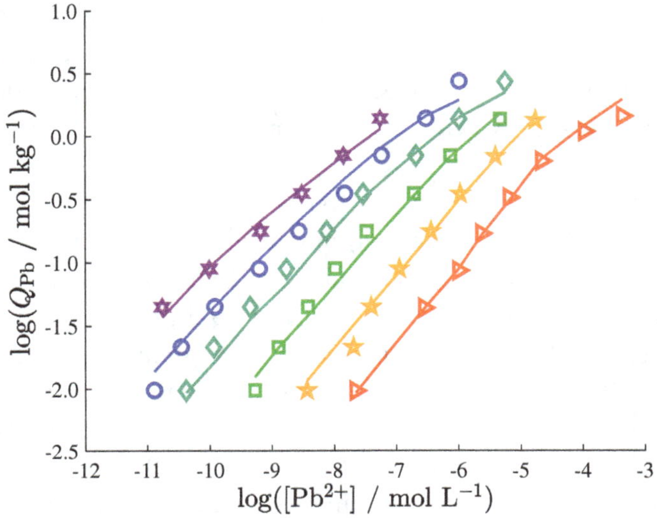

Figure 16. Isotherms with the loading of Pb^{2+} to humic acid. Red triangle markers: pH 4; orange 5-pointed star markers: pH 5; green square markers: pH 6; turquoise diamond markers: pH 7; blue circle markers: pH 8 and purple 6-pointed star markers: pH 9. Continuous line for the fitting with NICA model. Adapted from Ref. 31.

In this study, electrochemical results were compared to those predicted with the Windermere Humic Acid Model (WHAM).

The binding isotherms in presence of purified peat humic acid from the Mogi river region of Ribeirao Preto (Sao Paulo State, Brazil) have also been obtained for indium.[40,66]

AGNES technique has also been applied to the characterisation of soil extracts[56] from four different soil types (river clay, loam, cover sand, and reclaimed peat) from grassland fields in the Netherlands. After an extraction with $CaCl_2$ (soil solution ratio 1:10), AGNES (and also DMT) analysis were performed for the determination of free Zn^{2+} concentration. AGNES was also used in the context of a refined electrostatic model for nanoparticles.[67,68]

VII. Application of AGNES in Natural Samples

Another important field of application of AGNES technique is the study of metal speciation in natural waters. Different types of samples have been considered depending on their salinity (high, intermediate, and low), *i.e.*, sea, estuarine, and river waters. In many instances, the ionic strength of the sample (μ in mol L^{-1}) has been estimated with Russell's equation,[69] $\mu = 1.6 \times 10^{-5} \times EC$, where EC (μS cm^{-1}) stands for the electrical conductivity.

The free Zn concentration was found[70] in the range of 1–3 nmol L^{-1} (25% of the total Zn concentration) in two coastal Mediterranean seawaters from Tarragona and Castelldefelds (near Barcelona) in North-Eastern Spain, whose conductivities were 48.5 and 49.5 mS cm^{-1}. An essential improvement for this determination was the development of a new kind of blank, the shifted blank, which can be obtained in the same water sample (in the presence of the analyte), but with a suitable selection of the potentials towards a range where no faradaic current is detected.

Diaz de Alba and coworkers[71] measured the free Pb^{2+} concentration with AGNES in Atlantic Ocean sea water from Faro Beach (southern Portugal), with EC = 49.5 mS cm^{-1}, spiked with 6×10^{-7} mol L^{-1} Pb at pH values of 2.7, 5.0, 6.0, 7.0, and 8.6. These values compared well with the ones predicted by VM 3.0.

Free Zn concentration determined with AGNES across the Tamar river estuary (England), covering a factor of 300 in salinity,[72] ranged from 2 to 26 nmol L^{-1} and compared well (except for one sample) with the CLE-AdCSV (Competitive Ligand Exchange Adsorptive Cathodic Stripping Voltammetry) results.

River waters (typically with low ionic strength) have also been tackled with AGNES. Free Zn concentration was determined in Besòs river water with EC = 1.912 mS cm^{-1},[73] in Rhine river with EC = 0.720 mS cm^{-1},[56] in rivers Garonne, Gave de Cauterets and Gave de Pau in the Pyrenees, with EC ranging from 0.078 to 0.226 mS cm^{-1}, and with sampling points downstream from mining areas.[74] Free Cd concentration was measured in two natural water samples from Itapanhau and Sorocabinha rivers in Sao Paulo State (Brazil), with EC = 0.189 and 1.043 mS cm^{-1}, respectively, spiked with cadmium.[75]

A key point in the measurement of freshwater samples (and other complex matrices) has been the implementation of a novel device to purge with a mixture of N_2/CO_2 to remove the interfering oxygen while

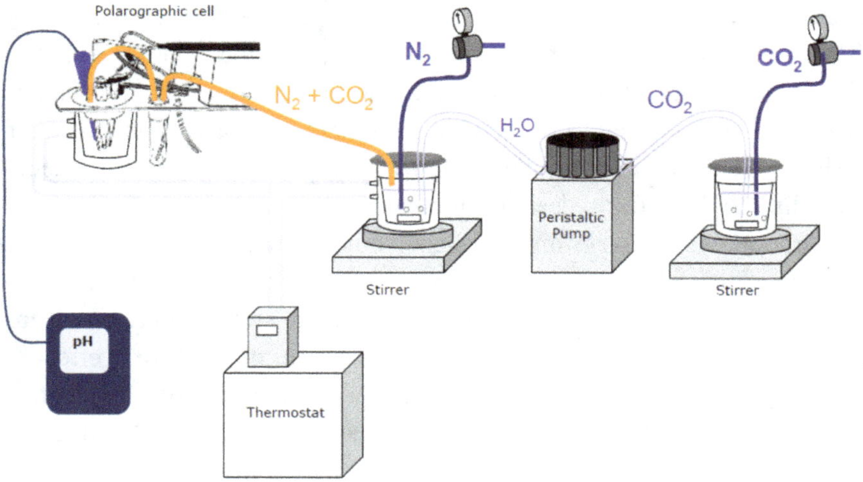

Figure 17. Device to control pH with a mixture of N_2 and CO_2. Adapted from Ref. 61.

maintaining the pH fixed,[13,28,56,61,73,76–78] as the typical purge with only N_2 eliminates CO_2 and induces a change in pH (see Figure 17).

Free Zn^{2+} levels were measured with AGNES on-site using an SPE and without purging in the Gave de Cauterets river in Soulom (France).[74]

The measurement of free metal concentration of freshwater samples overcame the difficulty of the lack of validity of the DPP at low ionic strength (the DPP peak potential usually measured to determine the relationship between gain and potential through Eq. 2). A way through consists in performing the DPP at a higher ionic strength and apply a correction (essentially equivalent to Eq. 8[27]) to obtain the gains at the low ionic strength.

VIII. Application of AGNES in Dispersions of Nanomaterials

The industrial production of metal-based nanomaterials is currently increasing due to their wide range of practical applications in ceramics, paints, coatings, personal care products, rubber manufacturing, fertilizers, biocides, *etc.*[79] Accordingly, there is a growing concern on the health and safety issues related with their eventual release into the environment.

Physicochemical characterization of the aqueous dispersions of nanomaterials (NMs) is key to understand the potentially hazardous effects of NMs, evaluate their risks, and improve the safety of novel products according to the safe-by-design principles.

Among all the relevant physicochemical endpoints for NM risk assessment, dissolution in aqueous media is one of the most important, for several reasons. Firstly, it determines the release of soluble ions, which are arguably the main chemical species responsible for the NM toxic effect.[80] Secondly, dissolution processes regulate mostly the effective concentration and size of particles, and thus the fate and behavior of NMs in environmental and biological compartments.

There are several features of NMs that affect specifically their rates and extents of dissolution. These features are responsible for the differences in reactivity of NMs with respect to their bulk counterparts. Namely, NMs typically display very small particle sizes and, concomitantly, large specific surface areas. This entails high solid-liquid mass-transfer rates, as well as very low local curvature radii of the solid surface. The former effect implies larger dissolution rates, whereas the latter may lead to an increased interfacial energy and, thus, an enhancement of solubility according to the Ostwald-Freundlich equation.[81] Moreover, the primary nanoparticles often show a varying degree of polydispersity in their size and shape. Additionally, these particles may suffer colloidal aggregation processes, leading to the formation of agglomerates and aggregates with a wide range of sizes and internal structures. Overall, the particle size distribution in NM dispersions is a dynamic property that typically shows a huge sensitivity to environmental variables (pH, temperature, ionic strength, *etc.*). Therefore, the effective surface areas and dissolution rates are usually a complex function of the experimental conditions.

The conventional method to investigate dissolution and solubility of a solid in a liquid matrix involves an initial solid-liquid separation step, such as (ultra)filtration, (ultra)centrifugation, *etc.*[82] In this context, one of the most relevant advantages of AGNES is the robustness in relation to the presence of suspended particles, which allows direct *in situ* probing of the liquid phase in the dispersion, thus avoiding the initial separation step. Besides, the standard thermostated and air-tight voltammetric cells ensure a suitable control over environmental variables (pH, temperature, partial pressure of CO_2, *etc.*) throughout the experiments. All this, together with the ability of sensing specifically and unambiguously the free ion species, makes AGNES a highly suitable technique to study the dissolution of NMs and/or the adsorption of concurrent ions on the NM interface in aqueous dispersions. The downside is that this technique is restricted to a limited number of possible metal ions, as discussed above. Most of the literature published to date deals with the dissolution of Zn and Cd-based NMs, plus several studies of the adsorption of Cd^{2+} and Pb^{2+} ions on other kinds of metallic and non-metallic NMs, as summarized below.

A. Zn-based Nanomaterials

The dissolution kinetics and solubility equilibria of uncoated ZnO nanoparticles dispersed in inert background electrolytes (KCl, KNO_3) was studied at 25 °C as a function of pH, primary particle size, and concentration.[83] The relatively short deposition times required for AGNES measurement of dissolved Zn^{2+} in these conditions (typically around 100 s) allowed the authors to record the ZnO dissolution rates (one point every 10 min). The experimental data were described using a reaction-diffusion model involving a single-step, reversible chemical reaction at the ZnO/water interface plus the diffusion of Zn^{2+} ions in the liquid boundary layer around the nanoparticles:

$$\mathrm{ZnO_s} + 2\mathrm{H}^+ \underset{k_{\mathrm{in}}}{\overset{k_{\mathrm{out}}}{\rightleftharpoons}} \mathrm{Zn}_{r_0}^{2+} \xrightarrow{\textit{diff}} \mathrm{Zn}^{2+} \tag{20}$$

where $Zn_{r_0}^{2+}$ and Zn^{2+} represent the free zinc ions at the ZnO surface and in the bulk solution, respectively. Note that water molecules are implicit. The expression of the free zinc(II) concentration at time t, $[Zn^{2+}]_t$, is given by Eq. 21,

$$\ln\left(1-\frac{\left[\mathrm{Zn}^{2+}\right]_t}{\left[\mathrm{Zn}^{2+}\right]_{\mathrm{eq}}}\right)=c-bt \tag{21}$$

where $[Zn^{2+}]_{eq}$ is the concentration of free Zn^{2+} ions at equilibrium, c is the value of the left hand side of the equation at $t = 0$, and b is defined as:

$$b = \frac{3c_{ZnO}}{\rho\, r_0}\left(\frac{k_{in} D_{Zn^{2+}}}{k_{in} r_0 + D_{Zn^{2+}}}\right) \tag{22}$$

where c_{ZnO} (kg m^{-3}) is the total mass concentration of ZnO in the dispersion, ρ (kg m^{-3}) is the effective density of the agglomerates (which is lower than the bulk ZnO density), r_0 (m) is the average radius of the agglomerates, k_{in} is the kinetic rate constant for the surface precipitation step, Eq. 20, and $D_{Zn^{2+}}$ is the diffusion coefficient of the free Zn^{2+} ions in the liquid boundary layer around the nanoparticles. In Eq. 21, $[Zn^{2+}]_{eq}$ depends on pH according to the solubility product reaction:

$$ZnO_{s,zincite} + H_2O \overset{K_{sp}}{\rightleftharpoons} Zn^{2+} + 2OH^- \tag{23}$$

The slope b in Eq. 21 depends on the total concentration of ZnO and the radius r_0 of the aggregates/agglomerates in the dispersion (not the primary particles). The former parameter may be safely assumed to remain constant during the dissolution process, whenever the initial amount of ZnO is much larger than the total solubility at equilibrium, which means that the equation is only valid for dissolution rates in solutions where the initial ZnO_s concentration is well above the saturation concentration. The latter parameter r_0 is dependent on the variation of the particle size distribution of the dispersion throughout the experiment. In fact, the addition of a well-dispersed, concentrated ZnO stock dispersion on a pH-buffered background salt solution at $t = 0$ leads to aggregation and dissolution processes taking place simultaneously and, thus, a decreasing value of b over time. Alternatively, the dissolution rate can be quantified through carefully designed experiments where the pH of a previously equilibrated ZnO dispersion is suddenly perturbed by the addition of a strong acid, which leads to an increase in $[Zn^{2+}]_{eq}$ as discussed below, while the average size of the agglomerates remains constant (as the ionic strength of the dispersion does not change).

The value of b can be related straightforwardly with the time $t_{1/2}$ necessary to reach a free Zn^{2+} concentration equal to one half the concentration at equilibrium (Eq. 24), *i.e.* $[Zn^{2+}]_{t_{1/2}} = 0.5 \times [Zn^{2+}]_{eq}$, starting from an initial state of $[Zn^{2+}]_{t=0} = 0$.

$$t_{1/2} = \frac{\ln 2}{b} \tag{24}$$

Figure 18 represents some theoretical values of $t_{1/2}$ calculated with this model, as a function of c_{ZnO} and r_0.

The values of $[Zn^{2+}]_{eq}$ measured as a function of pH and ionic strength can be described with the solubility product K_{sp} that corresponds to Eq. 23 above. The experimental values of K_{sp}, though, show an Ostwald-Freundlich dependence with the radius r of the primary ZnO particle.[83] Eventually, the following relationship (Eq. 25) is observed for $[Zn^{2+}]_{eq}$ as a function of size, pH and ionic strength of the medium:

$$\log\left(\frac{\left[Zn^{2+}\right]_{eq}}{\text{mol L}^{-1}}\right) = \frac{2\gamma\overline{V}}{RTr}\log e + \log\left(\frac{K_{sp}^{bulk}}{\gamma_{Zn^{2+}}K_w^2}\right) - 2\,\text{pH} \tag{25}$$

where γ is the $ZnO_{s,zincite}$/water interfacial energy (with a fitted value of 0.32 J m^{-2}), $\overline{V}$ is the molar volume of bulk $ZnO_{s,zincite}$, e is Euler's number,

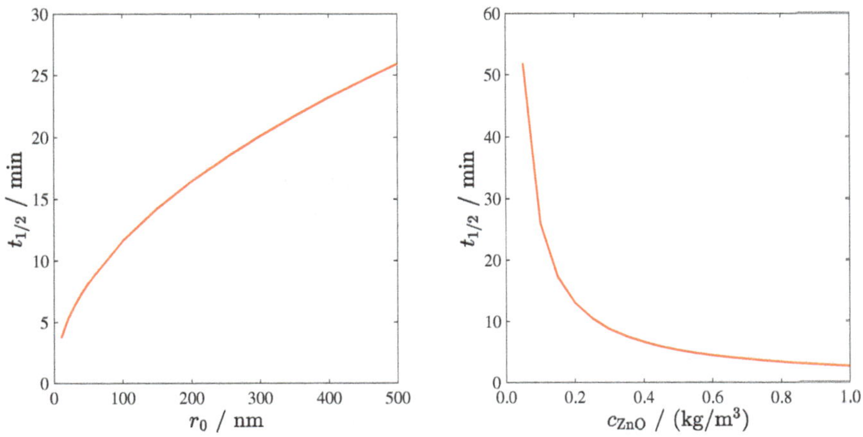

Figure 18. Theoretical (model) values of half-lifes ($t_{1/2}$) for the dissolution rates of ZnO nanoparticles in excess at 25 °C in 0.1 mol L^{-1} KCl as a function of c_{ZnO} for an aggregate radius r_0 = 500 nm (right) and values of $t_{1/2}$ as a function of r_0 for c_{ZnO} = 0.1 kg m^{-3} (left). Data calculated using fitted parameter values taken from Ref. 83.

$\gamma_{Zn^{2+}}$ is the activity coefficient for Zn^{2+} ions, K_w is the ionic product of water and K_{sp}^{bulk} is the solubility product of bulk zincite (fitted as $\log K_{sp}^{bulk} = -16.90 \pm 0.07$ at 25 °C). Theoretical values calculated using Eq. 25 are shown in Figure 19.

In practice, the effect of particle size on the solubility of ZnO is only noticeable below a diameter of 20 nm. Note that the dissolved concentration in equilibrium with ZnO NMs depends on the size of the primary particles r, whereas the rate of dissolution is controlled by the size of the agglomerates r_0.

As opposed to most sparingly soluble oxides, ZnO displays reverse (or retrograde) solubility because of the negative value of its enthalpy of dissolution: $\Delta H^0 = -88.7$ kJ mol^{-1} for Eq. 20.[84] This means that the values of $[Zn^{2+}]_{eq}$ increase as temperature decreases, while keeping the rest of environmental conditions constant (in particular, pH). This effect has also been confirmed by AGNES experiments (unpublished results).

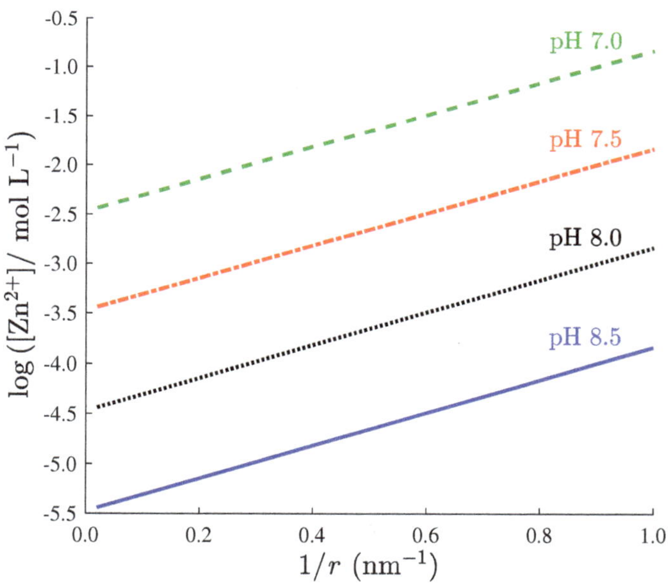

Figure 19. Theoretical values for the free Zn^{2+} concentration at equilibrium (log value) in dispersions of ZnO nanoparticles as a function of $1/r$ at different values of pH at 25 °C. Data calculated using fitted parameter values taken from Ref. 83.

In the case of ZnO NMs coated with surfactants, AGNES can still be used with an adequate design of the deposition conditions to avoid or minimize fouling of the hanging mercury drop electrode.[24] On the other hand, the presence of dissolved natural organic matter (humic acids) at concentrations as high as 0.45 g L^{-1} does not represent a noticeable interference for AGNES.[62] The adsorption of humic acid on ZnO NMs does not influence the free Zn^{2+} ion concentration at equilibrium with respect to pristine ZnO dispersed in media of the same composition and pH (unpublished results), in compliance with the solubility product reaction (Eq. 23).

Even though AGNES only provides a direct measurement of $[Zn^{2+}]$, the total Zn solubility can be derived indirectly from the mass balance relationship:

$$S=\left[\mathrm{Zn}^{2+}\right]_{\mathrm{eq}}+\left[\mathrm{Zn(OH)}^{+}\right]+\left[\mathrm{ZnL}^{2+}\right]+$$
$$+....=\left[\mathrm{Zn}^{2+}\right]\left\{1+K^{\mathrm{cond}}_{\mathrm{Zn(OH)}^{+}}\left[\mathrm{OH}^{-}\right]+K^{\mathrm{cond}}_{\mathrm{ZnL}^{2+}}\left[\mathrm{L}\right]+..\right\} \quad (26)$$

where $K^{\mathrm{cond}}_{\mathrm{Zn(OH)}^{+}}, K^{\mathrm{cond}}_{\mathrm{ZnL}^{2+}}$, *etc.* are the conditional formation constants for the corresponding inorganic (*e.g.* $Zn(OH)^{+}$) and organic complexes (*e.g.* ZnL^{2+}). This equation is useful whenever the medium composition is known, and the conditional values of the corresponding stability constants can be interpolated from available tabulated data. Alternatively, though, the equilibrium solubility of ZnO in a complex medium can be still obtained indirectly from addition experiments through a stepwise increase in the total Zn concentration. In fact, if excess of ligand conditions prevail, Eq. 26 leads to:

$$\log S=\log\left[\mathrm{Zn}^{2+}\right]_{\mathrm{eq}}+\log\left(1+K'\right) \quad (27)$$

where $K'=\Sigma_i K_i^{\mathrm{cond}}[\mathrm{L}_i]$ is the (overall) conditional formation constant corresponding to the medium composition (the summation extends over all ligands L_i either organic or inorganic, stoichiometric indexes are omitted for clarity). As a result, the conditional value of the ZnO solubility can be obtained from the intersection of the straight line log $[Zn^{2+}]_{eq}$ *vs.* $c_{T,Zn}$ (undersaturated solutions) and the constant value of log $[Zn^{2+}]_{eq}$ achieved at saturated conditions (see dashed lines in Figure 20).

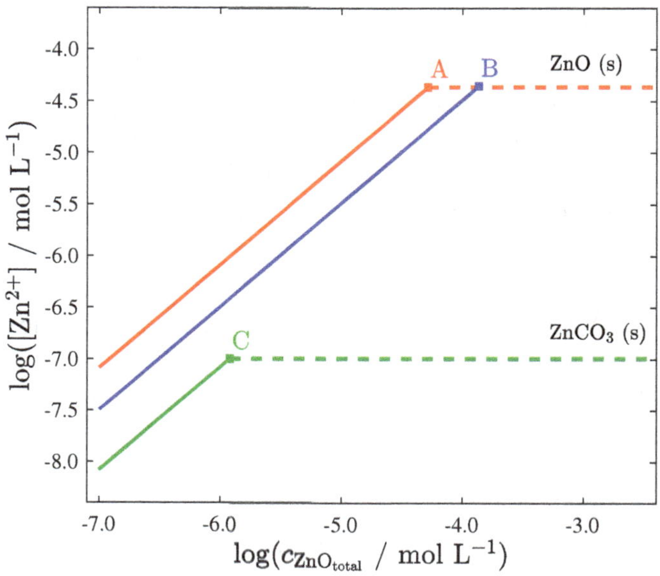

Figure 20. Theoretical values of the free Zn^{2+} concentration at equilibrium (log value) as a function of the total ZnO concentration in a background solution of 0.1 mol L^{-1} KCl at pH 8 and 25 °C in the absence of $CO_{2(g)}$ and organic ligands (red line A), in the presence of an organic ligand (blue line B), and in the presence of 5% $CO_{2(g)}$ (green lines C). Red line A: $K' = 0.2$ (only inorganic complexation due to Cl^- ions) and $S = 0.004$ g L^{-1}. Blue line B: $K' = 0.2 + 1.9 = 2.1$ (where 1.9 is the contribution of the organic ligand to the overall complexation constant of the medium) and $S = 0.011$ g L^{-1}. Green lines C: the presence of dissolved carbonate species induces precipitation of solid $ZnCO_3$ above 0.0001 g L^{-1} of total ZnO concentration.

This indirect procedure to obtain the ZnO solubility was tested in fluids of biological interest such as Dulbecco's Modified Eagle Medium (DMEM), a culture media for *in vitro* tests with mammalian cell lines,[77] and synthetic saliva.[78] In DMEM at 37 °C, 5% CO_2, and pH 7.68, the relatively high concentration of dissolved carbonates led to the coupled dissolution of ZnO and precipitation of (allegedly) smithsonite ($ZnCO_3$) above a total ZnO concentration of 5.5 mg L^{-1}. Conversely, in saliva at 37 °C and pH 6.80, the presence of dissolved phosphates led to the presumed precipitation of hopeite ($Zn_3(PO_4)_2{\cdot}4H_2O$) above 0.47 mg L^{-1} of ZnO. In both media, the experimentally determined values of the overall conditional formation constant K' were consistent with the thermodynamic data available. Most notably, the solubility determined in synthetic saliva by AGNES was in good agreement with the value measured by ICP-MS after ultrafiltration of the ZnO dispersions. However, the latter method proved to be less reproducible and more prone to bias due to inadequate control of environmental variables, such as temperature or CO_2 partial pressure, during the solid/liquid separation step.

AGNES has also been used to explore the dissolution behavior of ZnO nanoparticles at concentrations below saturation during chronic toxicity tests with *Daphnia magna*, demonstrating that all the ZnO was dissolved within 48 h of exposure.[80]

B. Other Metal Ions

There are several examples of the use of AGNES for the measurement of other free metal ions besides Zn^{2+} in the liquid phase solution of NM dispersions. For instance, the release of Cd^{2+} ions from CdTe/CdS quantum dots was investigated through this technique to determine the role of dissolved free ions on the bioaccumulation of Cd by *Chlamydomonas reinhardtii*[85] and to study the dissolution behavior of these nanoparticles in the presence of dissolved organic ligands.[86] The dissolution kinetics of In_2O_3 nanoparticles at different pH values (in 0.1 mol L^{-1} KNO_3) and in synthetic seawater (pH 8, where a free indium concentration in the amol L^{-1} range was determined) has been studied.[41]

The adsorption of trace metal ions on insoluble NMs in aqueous dispersions has also been characterized through AGNES: Cd^{2+} and Pb^{2+} interaction with carboxylated polystyrene latex,[42,60,87] Cd^{2+} interaction with TiO_2,[88,89] and Cd^{2+} and Pb^{2+} adsorption on CeO_2 (unpublished work).

IX. Application of AGNES for [In^{3+}] and [$Sb(OH)_3$] Determination

A. Trivalent Redox States

Trivalent redox state elements experience a large hydrolysis in aqueous media, so that the resulting free element concentration is extremely low at circumneutral pH values. This difficulty is additionally coupled with a limited solubility of some formed hydroxides or of the reduced element in the amalgam. Hence, AGNES analysis of elements with trivalent redox state needs a different approach to that of the divalent ones (Zn^{2+}, Cd^{2+}, Pb^{2+}). Indeed, one cannot rely on the application of Eq. 2 to find the gain Y, either because of irreversibility (in the case of indium) or of the overwhelming presence of complexes (in the case of antimony) at any workable pH. So, a strategy consists in computing the value of η_Q from the volume of mercury in the used electrode with Eq. 7. One performs a calibration (*i.e.* plot Q *vs.* $[In^{3+}]$ in the case of indium or plot Q *vs.* $[Sb(OH)_3]$ in the case of antimony, see below) at a fixed deposition potential E_{calib}. According to Eq. 6, the slope in the calibration can be interpreted as Y_{calib} times η_Q, where Y_{calib} is the gain corresponding to the potential E_{calib}. Once this correspondence (E_{calib}, Y_{calib}) has been established from the calibration,[34,40] any other gain can be computed for a given potential or *vice versa* by using Eq. 28.

$$E_j = E_{calib} - \frac{RT}{3F} \ln \frac{Y_j}{Y_{calib}} \tag{28}$$

B. Indium

Despite the described irreversibility[11,34,55] of the In^0/In^{3+} couple (see Section IV.E.3), equilibrium could be reached. In the case of HMDE, the rule of time for 1P at pH 3 was given by Eq. 19, see Figure 12. Comparison of AGNES free indium concentrations in NTA+In^{3+} mixtures with various speciation models (from sets of stability constants in the literature), allowed[34] to recommend those of Biver *et al.*[90] as the most reliable ones (see Figure 21).

To evaluate the solubility product of $In(OH)_3(s)$, $[In^{3+}]$ was measured at various pH values in solution (see Figure 22) where the precipitate was present.[48] $[In^{3+}]$ decreases by a factor of 1000 for each pH unit increase. At pH 5.6, the expected $[In^{3+}]$ is around 15×10^{-12} mol L^{-1}, so, a gain around 5×10^6 is required. The rule for AGNES-2P for HMDE (not accounting for the contribution of any complex), Eq. 16, indicates a

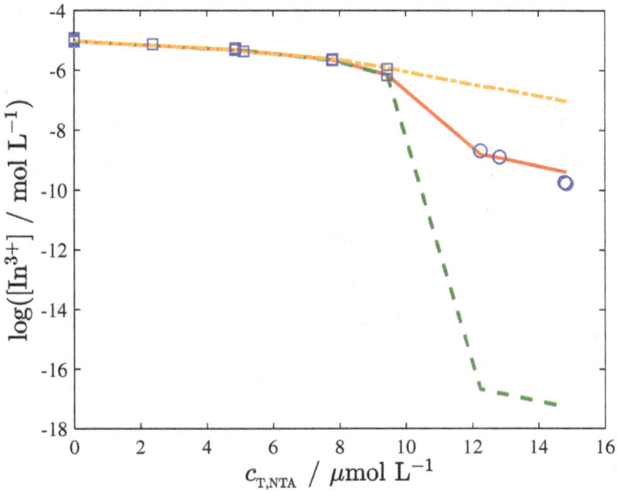

Figure 21. Indium-NTA speciation at pH = 3; $c_{T,In} = 10^{-5}$ mol L^{-1}, [KNO_3] = 0.1 mol L^{-1}, T = 25.0 °C, Y from 2 to 10^5, deposition times from 200 s to 2000 s. Blue square markers: AGNES-1P; blue circle markers: AGNES-2P. Green dashed line: Harris *et al.* model.[96] Red continuous line: Biver *et al.* model.[90] Dotted-dashed orange line: NIST 46.6 model (default database in Visual MINTEQ[26]). Adapted from Ref. 34.

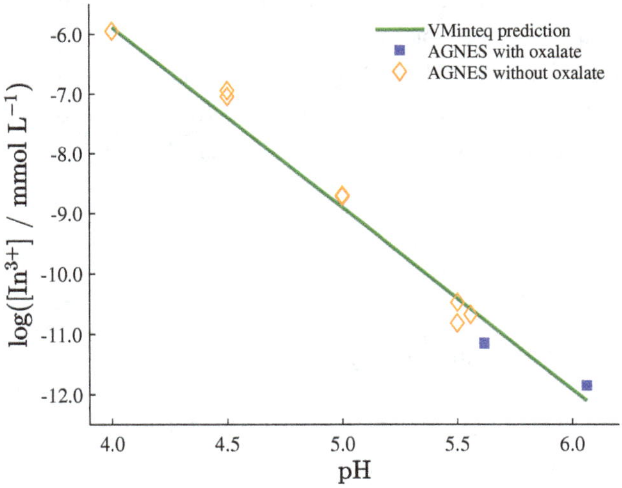

Figure 22. Indium solubility, at various pH-values, given by the free indium concentration in equilibrium with precipitated $In(OH)_3$ ruled by its K_{sp}. For the higher pH-values, a strategy of adding oxalate was used to enhance the flux and reduce deposition time, while not disturbing [In^{3+}]. [KNO_3] = 0.1 mol L^{-1}, T = 25.0 °C. Adapted from Ref. 48.

deposition time of around 50 days. The used strategy was to dramatically enhance the labile complexes present in the medium by adding oxalate, which does not alter $[In^{3+}]$ as long as there is the solid phase $In(OH)_3(s)$ and fixed pH. The approximate helping factor (see Section IV.E.2) was around 10 000, so that the estimated time (with oxalate contributions) was 5×10^6 /10 000 = 500 s. Experiments showed that just $t_{1,a} = 500$ s was enough to reach the desired equilibration.[48]

Application of the RDE at pH 2.6 led to Eq. 29, which is similar to that for divalent redox state elements (see Eq. 15).[40]

$$t_1 - t_w = 0.005 \times Y \tag{29}$$

C. Antimony

Due to the uncertain existence of the species Sb^{3+} in aqueous solution, AGNES is calibrated with respect to $[Sb(OH)_3]$. Indeed, when AGNES conditions are reached, one can choose to calibrate with respect to any species of non-negligible concentration among those in equilibrium, because fixed relationships ruled by stability constants and known concentrations of other species will eventually hold between their concentrations and $[M^0]$ at a fixed potential.

The gains relating $[Sb^0]$ and $[Sb(OH)_3]$ are labeled with a prime in Eq. 30 to highlight the fact that this gain is also a function of pH. Expressions to relate the Y' at different pH values have been derived.[91]

$$Y' \equiv \frac{\left[Sb^0\right]}{\left[Sb(OH)_3\right]} \tag{30}$$

The fundamental equation is just:

$$Q = \eta_Q Y' \left[Sb(OH)_3\right] \tag{31}$$

The derived rule of time (at pH 2, for HMDE) is given by Eq. 32.

$$t_1 - t_w = 30 \times Y' \tag{32}$$

Sb speciation was checked in titrations with oxalate.

X. Conclusions

AGNES has successfully enabled to tackle the speciation of a range of elements, including Zn(II), Cd(II), Pb(II), Sb(III), and In(III). Unpublished results also confirm the possibility of using AGNES with Sn(II) and Ga(III).[92] A future candidate element for AGNES is Tl. Solid electrodes might extend AGNES application beyond amalgamating elements, but non-mercury electrodes are more difficult to handle.

By providing a direct and robust measurement of the concentration of the free ion or a species related to it, AGNES can be used to discriminate between competing speciation models in synthetic or complex natural media, as shown in the cases of In-NTA (see Figure 21) and of Zn-GSH (see Figure 15).

AGNES has been applied in a wide range of systems, spanning from natural seawater to wine, nanoparticle dispersions or fetal bovine serum. Indeed, AGNES is not hindered by turbidity of the sample and does not require any previous separation step. *On-site* measurement has been successfully achieved with the use of screen-printed electrodes, despite the low ionic strengths found in these freshwaters.

Further work on interferences, miniaturization, biological systems, *etc.* would be helpful in order to develop the applicability of this technique in the future.

XI. Acknowledgments

Support from the Spanish Minister of Science and Innovation MCIN/AEI/10.13039/501100011033 is gratefully acknowledged (projects PID2019-107033GB-C21, PID2020-117910GB-C21 and and PID2022-140312NB-C21). A large number of co-workers of the authors of this chapter (mostly contributing to referenced articles) is acknowledged.

XII. References

1. Bakker, E.; Pretsch, E. Modern Potentiometry. *Angew. Chem., Int. Ed. Engl.* **2007**, *46*, 5660–5668.
2. Fortin, C.; Campbell, P. G. C. An Ion-exchange Technique for Free-metal Ion Measurements (Cd^{2+}, Zn^{2+}): Applications to Complex Aqueous Media. *Int. J. Environ. Anal. Chem.* **1998**, *72*, 173–194.

3. Pesavento, M.; Biesuz, R.; Alberti, G. Determination of the Metal Complexation Capacity of Natural Waters by Ligand Titration in the Presence of Complexing Resins. *J. Phys. IV* **2003**, *107*, 1045–1048.
4. Temminghoff, E. J. M.; Plette, A. C. C.; van Eck, R.; van Riemsdijk, W. H. Determination of the Chemical Speciation of Trace Metals in Aqueous Systems by the Wageningen Donnan Membrane Technique. *Anal. Chim. Acta* **2000**, *417*, 149–157.
5. Vera, R.; Fontas, C.; Galceran, J.; Serra, O.; Antico, E. Polymer Inclusion Membrane to Access Zn Speciation: Comparison with Root Uptake. *Sci. Total Environ.* **2018**, *622*, 316–324.
6. Davison, W. (Ed.) *Diffusive Gradients in Thin-Films for Environmental Measurements*; Cambridge University Press: Cambridge, UK, 2016.
7. Bard, A. J.; Faulkner, L. R. *Electrochemical Methods. Fundamentals and Applications*, 2nd ed.; John Wiley & Sons, Inc.: New York, 2001.
8. Galceran, J.; Cecília, J.; Salvador, J.; Monné, J.; Torrent, M.; Companys, E.; Puy, J.; Garcés, J. L.; Mas, F. Voltammetric Currents for any Ligand-to-metal Concentration Ratio in Fully Labile Metal-macromolecular Complexation. Easy Computations, Analytical Properties of the Currents and a Graphical Method to Estimate the Stability Constant. *J. Electroanal. Chem.* **1999**, *472*, 42–52.
9. Companys, E.; Galceran, J.; Pinheiro, J. P.; Puy, J.; Salaün, P. A Review on Electrochemical Methods for Trace Metal Speciation in Environmental Media. *Curr. Opin. Electrochem.* **2017**, *3*, 144–162.
10. van Leeuwen, H. P.; Town, R. M. Stripping Chronopotentiometry at Scanned Deposition Potential (SSCP). Part 1. Fundamental Features. *J. Electroanal. Chem.* **2002**, *536*, 129–140.
11. Galceran, J.; Companys, E.; Puy, J.; Pinheiro, J. P.; Rotureau, E. AGNES in Irreversible Systems: The Indium Case. *J. Electroanal. Chem.* **2021**, *901*, 115750.
12. Galceran, J.; Puy, J. Interpretation of Diffusion Gradients in Thin Films (DGT) Measurements: A Systematic Approach. *Environ. Chem.* **2015**, *12*, 112–122.
13. Rosales-Segovia, K.; Sans-Duñó, J.; Companys, E.; Puy, J.; Alcalde, B.; Antico, E.; Fontas, C.; Galceran, J. Effective Concentration Signature of Zn in a Natural Water Derived from Various Speciation Techniques. *Sci. Total Environ.* **2022**, *806*, 151201.
14. Companys, E.; Puy, J.; Torrent, M.; Galceran, J.; Salvador, J.; Garcés, J. L.; Mas, F. Binding Curve from Normalised Limiting Currents of Labile Heterogeneous Metal-macromolecular Systems. The Case of Cd/humic Acid. *Electroanalysis* **2003**, *15*, 452–459.
15. Pinheiro, J. P.; Galceran, J.; Rotureau, E.; Companys, E.; Puy, J. Full Wave Analysis of Stripping Chronopotentiometry at Scanned Deposition Potential (SSCP): Obtaining Binding Curves in Labile Heterogeneous Macromolecular Systems for Any Metal-to-ligand Ratio. *J. Electroanal. Chem.* **2020**, *873*, 114436.
16. Serrano, N.; Díaz-Cruz, J. M.; Ariño, C.; Esteban, M.; Puy, J.; Companys, E.; Galceran, J.; Cecilia, J. Full-wave Analysis of Stripping Chronopotentiograms at Scanned Deposition Potential (SSCP) as a Tool for Heavy Metal Speciation:

Theoretical Development and Application to Cd(II)-phthalate and Cd(II)-iodide Systems. *J. Electroanal. Chem.* **2007**, *600*, 275–284.

17. Salaün, P.; Buffle, J. Integrated Microanalytical System Coupling Permeation Liquid Membrane and Voltammetry for Trace Metal Speciation. Theory and Applications. *Anal. Chem.* **2004**, *76*, 31–39.
18. Sigg, L.; Black, F.; Buffle, J.; Cao, J.; Cleven, R.; Davison, W.; Galceran, J.; Gunkel, P.; Kalis, E.; Kistler, D.; Martin, M.; Noel, S.; Nur, Y.; Odzak, N.; Puy, J.; van Riemsdijk, W. H.; Temminghoff, E.; Tercier-Waeber, M. L.; Toepperwien, S.; Town, R. M.; Unsworth, E.; Warnken, K. W.; Weng, L. P.; Xue, H. B.; Zhang, H. Comparison of Analytical Techniques for Dynamic Trace Metal Speciation in Natural Freshwaters. *Environ. Sci. Technol.* **2006**, *40*, 1934–1941.
19. Companys, E.; Galceran, J.; Puy, J.; Sedo, M.; Vera, R.; Antico, E.; Fontas, C. Comparison of Different Speciation Techniques to Measure Zn Availability in Hydroponic Media. *Anal. Chim. Acta* **2018**, *1035*, 32–43.
20. Companys, E.; Cecília, J.; Codina, G.; Puy, J.; Galceran, J. Determination of the Concentration of Free Zn^{2+} with AGNES Using Different Strategies to Reduce the Deposition Time. *J. Electroanal. Chem.* **2005**, *576*, 21–32.
21. Galceran, J.; Companys, E.; Puy, J.; Cecília, J.; Garcés, J. L. AGNES: A New Electroanalytical Technique for Measuring Free Metal Ion Concentration. *J. Electroanal. Chem.* **2004**, *566*, 95–109.
22. Pesavento, M.; Alberti, G.; Biesuz, R. Analytical Methods for Determination of Free Metal Ion Concentration, Labile Species Fraction and Metal Complexation Capacity of Environmental Waters: A Review. *Anal. Chim. Acta* **2009**, *631*, 129–141.
23. Feldmann, J.; Salaün, P.; Lombi, E. Critical Review Perspective: Elemental Speciation Analysis Methods in Environmental Chemistry — Moving Towards Methodological Integration. *Environ. Chem.* **2009**, *6*, 275–289.
24. Galceran, J.; Lao, M.; David, C.; Companys, E.; Rey-Castro, C.; Salvador, J.; Puy, J. The Impact of Electrodic Adsorption on Zn, Cd or Pb Speciation Measurements with AGNES. *J. Electroanal. Chem.* **2014**, *722–723*, 110–118.
25. Lopez-Solis, L.; Companys, E.; Puy, J.; Blindauer, C. A.; Galceran, J. Direct Determination of Free Zn Concentration in Samples of Biological Interest. *Anal. Chim. Acta* **2022**, *1229*.
26. Gustafsson, J. P. Visual MINTEQ Version 3.1. 2016. https://vminteq.lwr.kth.se/download/
27. Aguilar, D.; Parat, C.; Galceran, J.; Companys, E.; Puy, J.; Authier, L.; Potin-Gautier, M. Determination of Free Metal Ion Concentrations with AGNES in Low Ionic Strength Media. *J. Electroanal. Chem.* **2013**, *689*, 276–283.
28. Chito, D.; Galceran, J.; Companys, E.; Puy, J. Determination of the Complexing Capacity of Wine for Zn Using the Absence of Gradients and Nernstian Equilibrium Stripping Technique. *J. Agric. Food. Chem.* **2013**, *61*, 1051–1059.
29. Aguilar, D.; Galceran, J.; Companys, E.; Puy, J.; Parat, C.; Authier, L.; Potin-Gautier, M. Non-purged Voltammetry Explored with AGNES. *Phys. Chem. Chem. Phys.* **2013**, *15*, 17510–17521.

30. Pernet-Coudrier, B.; Companys, E.; Galceran, J.; Morey, M.; Mouchel, J. M.; Puy, J.; Ruiz, N.; Varrault, G. Pb-binding to Various Dissolved Organic Matter in Urban Aquatic Systems: Key Role of the Most Hydrophilic Fraction. *Geochim. Cosmochim. Acta* **2011**, *75*, 4005–4019.
31. Puy, J.; Galceran, J.; Huidobro, C.; Companys, E.; Samper, N.; Garcés, J. L.; Mas, F. Conditional Affinity Spectra of Pb^{2+}-Humic Acid Complexation from Data Obtained with AGNES. *Environ. Sci. Technol.* **2008**, *42*, 9289–9295.
32. Companys, E.; Naval-Sanchez, M.; Martinez-Micaelo, N.; Puy, J.; Galceran, J. Measurement of Free Zinc Concentration in Wine with AGNES. *J. Agric. Food. Chem.* **2008**, *56*, 8296–8302.
33. van Leeuwen, H. P.; Town, R. M. Elementary Features of Depletive Stripping Chronopotentiometry. *J. Electroanal. Chem.* **2002**, *535*, 1–9.
34. Tehrani, M. H.; Companys, E.; Dago, A.; Puy, J.; Galceran, J. Free Indium Concentration Determined with AGNES. *Sci. Total Environ.* **2018**, *612*, 269–275.
35. Parat, C.; Aguilar, D.; Authier, L.; Potin-Gautier, M.; Companys, E.; Puy, J.; Galceran, J. Determination of Free Metal Ion Concentrations Using Screen-Printed Electrodes and AGNES with the Charge as Response Function. *Electroanalysis* **2011**, *23*, 619–627.
36. Galceran, J.; Chito, D.; Martinez-Micaelo, N.; Companys, E.; David, C.; Puy, J. The Impact of High Zn^0 Concentrations on the Application of AGNES to Determine Free Zn(II) Concentration. *J. Electroanal. Chem.* **2010**, *638*, 131–142.
37. Town, R. M.; van Leeuwen, H. P. Fundamental Features of Metal Ion Determination by Stripping Chronopotentiometry. *J. Electroanal. Chem.* **2001**, *509*, 58–65.
38. Parat, C.; Authier, L.; Aguilar, D.; Companys, E.; Puy, J.; Galceran, J.; Potin-Gautier, M. Direct Determination of Free Metal Concentration by Implementing Stripping Chronopotentiometry as Second Stage of AGNES. *Analyst* **2011**, *136*, 4337–4343.
39. Huidobro, C.; Companys, E.; Puy, J.; Galceran, J.; Pinheiro, J. P. The Use of Microelectrodes with AGNES. *J. Electroanal. Chem.* **2007**, *606*, 134–140.
40. Rotureau, E.; Pla-Vilanova, P.; Galceran, J.; Companys, E.; Pinheiro, J. P. Towards Improving the Electroanalytical Speciation Analysis of Indium. *Anal. Chim. Acta* **2019**, *1052*, 57–64.
41. Rosales-Segovia, K.; Companys, E.; Puy, J.; Galceran, J. Release of Indium from In_2O_3 Nanoparticles nanoparticles in model solutions and synthetic seawater, *Sci. Total Environ.* **2024**, *926*, 171784.
42. Rocha, L. S.; Companys, E.; Galceran, J.; Carapuca, H. M.; Pinheiro, J. P. Evaluation of Thin Mercury Film Rotating Disc Electrode to Perform Absence of Gradients and Nernstian Equilibrium Stripping (AGNES) Measurements. *Talanta* **2010**, *80*, 1881–1887.
43. Rocha, L. S.; Galceran, J.; Puy, J.; Pinheiro, J. P. Determination of the Free Metal Ion Concentration Using AGNES Implemented with Environmentally Friendly Bismuth Film Electrodes. *Anal. Chem.* **2015**, *87*, 6071–6078.

44. Domingos, R. F.; Carreira, S.; Galceran, J.; Salaün, P.; Pinheiro, J. P. AGNES at Vibrated Gold Microwire Electrode for the Direct Quantification of Free Copper Concentrations. *Anal. Chim. Acta* **2016**, *920*, 29–36.
45. Alberti, G.; Biesuz, R.; Huidobro, C.; Companys, E.; Puy, J.; Galceran, J. A Comparison Between the Determination of Free Pb(II) by Two Techniques: Absence of Gradients and Nernstian Equilibrium Stripping and Resin Titration. *Anal. Chim. Acta* **2007**, *599*, 41–50.
46. Galceran, J.; Puy, J.; Salvador, J.; Cecília, J.; van Leeuwen, H. P. Voltammetric Lability of Metal Complexes at Spherical Microelectrodes with Various Radii. *J. Electroanal. Chem.* **2001**, *505*, 85–94.
47. Puy, J.; Galceran, J. Theoretical Aspects of Dynamic Metal Speciation with Electrochemical Techniques. *Curr. Opin. Electrochem.* **2017**, *1*, 80–87.
48. Tehrani, M. H.; Companys, E.; Dago, A.; Puy, J.; Galceran, J. New Methodology to Measure Low Free Indium(III) Concentrations Based on the Determination of the Lability Degree of Indium Complexes. Assessment of $In(OH)_3$ Solubility Product. *J. Electroanal. Chem.* **2019**, *847*, 113185.
49. Galceran, J.; Puy, J.; Salvador, J.; Cecília, J.; Mas, F.; Garcés, J. L. Lability and Mobility Effects on Mixtures of Ligands Under Steady-state Conditions. *Phys. Chem. Chem. Phys.* **2003**, *5*, 5091–5100.
50. Salvador, J.; Garcés, J. L.; Galceran, J.; Puy, J. Lability of a Mixture of Metal Complexes Under Steady-state Planar Diffusion in a Finite Domain. *J. Phys. Chem. B* **2006**, *110*, 13661–13669.
51. Salvador, J.; Puy, J.; Cecilia, J.; Galceran, J. Lability of Complexes in Steady State Finite Planar Diffusion. *J. Electroanal. Chem.* **2006**, *588*, 303–313.
52. Salvador, J.; Puy, J.; Galceran, J.; Cecilia, J.; Town, R. M.; van Leeuwen, H. P. Lability Criteria for Successive Metal Complexes in Steady-state Planar Diffusion. *J. Phys. Chem. B* **2006**, *110*, 891–899.
53. Salvador, J.; Garcés, J. L.; Companys, E.; Cecilia, J.; Galceran, J.; Puy, J.; Town, R. M. Ligand Mixture Effects in Metal Complex Lability. *J. Phys. Chem. A* **2007**, *111*, 4304–4311.
54. Sans-Duñó, J.; Cecilia, J.; Galceran, J.; Puy, J. Availability of Metals to DGT Devices with Different Configurations. The Case of Sequential Ni Complexation. *Sci. Total Environ.* **2021**, *779*, 146277.
55. Town, R. M.; Duval, J. F. L.; van Leeuwen, H. P. Electrochemical Activity of Various Types of Aqueous In(III) Species at a Mercury Electrode. *J. Solid State Electrochem.* **2020**, *24*, 2807–2818.
56. Chito, D.; Weng, L.; Galceran, J.; Companys, E.; Puy, J.; van Riemsdijk, W. H.; van Leeuwen, H. P. Determination of free Zn^{2+} Concentration in Synthetic and Natural Samples with AGNES (Absence of Gradients and Nernstian Equilibrium Stripping) and DMT (Donnan Membrane Technique). *Sci. Total Environ.* **2012**, *421–422*, 238–244.

57. Lopez-Solis, L.; Galceran, J.; Puy, J.; Companys, E. Absence of Gradients and Nernstian Equilibrium Stripping (AGNES): An Electroanalytical Technique for Chemical Speciation: A Tutorial Review. *Chemosensors* **2022**, *10*, 351.
58. Pettit, L. D.; Powell, K. J. Stability Constants Database. SolEq Solution Equilibria: Principles and Applications. Academic Software. Otley, Yorks, UK. 2001. https://www.acadsoft.co.uk/
59. Puigdomenech, I. Medusa: Make Equilibrium Diagrams Using Sophisticated Algorithms. Royal Institute of Technology KTH, Stockholm, Sweden. 2001. https://www.kth.se/che/medusa
60. Domingos, R. F.; Huidobro, C.; Companys, E.; Galceran, J.; Puy, J.; Pinheiro, J. P. Comparison of AGNES (Absence of Gradients and Nernstian Equilibrium Stripping) and SSCP (Scanned Stripping Chronopotenciometry) for Trace Metal Speciation Analysis. *J. Electroanal. Chem.* **2008**, *617*, 141–148.
61. Lao, M.; Dago, A.; Serrano, N.; Companys, E.; Puy, J.; Galceran, J. Free Zn^{2+} Determination in Systems with Zn-Glutathione. *J. Electroanal. Chem.* **2015**, *756*, 207–211.
62. Companys, E.; Puy, J.; Galceran, J. Humic Acid Complexation to Zn and Cd Determined with the New Electroanalytical Technique AGNES. *Environ. Chem.* **2007**, *4*, 347–354.
63. Janot, N.; Pinheiro, J. P.; Botero, W. G.; Meeussen, J. C.; Groenenberg, J. E. PEST-ORCHESTRA, a Tool for Optimising Advanced Ion-binding Model Parameters: Derivation of NICA-Donnan Model Parameters for Humic Substances Reactivity. *Environ. Chem.* **2017**, *14*, 31–38.
64. Meeussen, J. C. L. ORCHESTRA: An Object-Oriented Framework for Implementing Chemical Equilibrium Models. *Environ. Sci. Technol.* **2003**, *37*, 1175–1182.
65. Chen, W. B.; Gueguen, C.; Smith, D. S.; Galceran, J.; Puy, J.; Companys, E. Metal (Pb, Cd, and Zn) Binding to Diverse Organic Matter Samples and Implications for Speciation Modeling. *Environ. Sci. Technol.* **2018**, *52*, 4163–4172.
66. Rotureau, E.; Pinheiro, J. P.; Duval, J. F. L. On the Evaluation of the Intrinsic Stability of Indium-Nanoparticulate Organic Matter Complexes. *Colloid Surf. A-Physicochem. Eng. Asp.* **2022**, *645*.
67. Town, R. M.; van Leeuwen, H. P. Intraparticulate Speciation Analysis of Soft Nanoparticulate Metal Complexes. The Impact of Electric Condensation on the Binding of $Cd^{2+}/Pb^{2+}/Cu^{2+}$ by humic acids. *Phys. Chem. Chem. Phys.* **2016**, *18*, 10049–10058.
68. Town, R. M.; van Leeuwen, H. P. Metal Ion-humic acid Nanoparticle Interactions: Role of Both Complexation and Condensation Mechanisms. *Phys. Chem. Chem. Phys.* **2016**, *18*, 18024–18032.
69. In *Standard methods for the Examination of Water and Wastewater*, Clesceri, L. S.; Greenberg, A. E.; Eaton, A. D., Eds.; American Public Health Association, American Water Works Association, and Water Environment Federation: Washington DC, 1998, pp. 2–30 to 32–33.

70. Galceran, J.; Huidobro, C.; Companys, E.; Alberti, G. AGNES: A Technique for Determining the Concentration of Free Metal Ions. The case of Zn(II) in coastal Mediterranean seawater. *Talanta* **2007**, *71*, 1795–1803.
71. Diaz-de-Alba, M.; Galindo-Riano, M. D.; Pinheiro, J. P. Lead Electrochemical Speciation Analysis in Seawater Media by Using AGNES and SSCP techniques. *Environ. Chem.* **2014**, *11*, 137–149.
72. Pearson, H. B. C.; Galceran, J.; Companys, E.; Braungardt, C.; Worsfold, P.; Puy, J.; Comber, S. Absence of Gradients and Nernstian Equilibrium Stripping (AGNES) for the Determination of [Zn^{2+}] in Estuarine Waters. *Anal. Chim. Acta* **2016**, *912*, 32–40.
73. Zavarise, F.; Companys, E.; Galceran, J.; Alberti, G.; Profumo, A. Application of the New Electroanalytical Technique AGNES for the Determination of Free Zn Concentration in River Water. *Anal. Bioanal. Chem.* **2010**, *397*, 389–394.
74. Parat, C.; Authier, L.; Castetbon, A.; Aguilar, D.; Companys, E.; Puy, J.; Galceran, J.; Potin-Gautier, M. Free Zn^{2+} Determination in Natural Freshwaters of the Pyrenees: Towards on-site Measurements with AGNES. *Environ. Chem.* **2015**, *12*, 329–337.
75. Monteiro, A. C. S.; Parat, C.; Rosa, A. H.; Pinheiro, J. P. Towards Field Trace Metal Speciation Using Electroanalytical Techniques and Tangential Ultrafiltration. *Talanta* **2016**, *152*, 112–118.
76. Pei, J. H.; Tercier-Waeber, M. L.; Buffle, J. Simultaneous Determination and Speciation of zinc, Cadmium, Lead, and Copper in Natural Water with Minimum Handling and Artifacts, by Voltammetry on a Gel-integrated Microelectrode Array. *Anal. Chem.* **2000**, *72*, 161–171.
77. Mu, Q. S.; David, C. A.; Galceran, J.; Rey-Castro, C.; Krzeminski, L.; Wallace, R.; Bamiduro, F.; Milne, S. J.; Hondow, N. S.; Brydson, R.; Vizcay-Barrena, G.; Routledge, M. N.; Jeuken, L. J. C.; Brown, A. P. Systematic Investigation of the Physicochemical Factors That Contribute to the Toxicity of ZnO Nanoparticles. *Chem. Res. Toxicol.* **2014**, *27*, 558–567.
78. David, C. A.; Galceran, J.; Quattrini, F.; Puy, J.; Rey-Castro, C. Dissolution and Phosphate-induced Transformation of ZnO Nanoparticles in Synthetic Saliva Probed by AGNES Without Previous Solid-liquid Separation. Comparison with UF-ICP-MS. *Environ. Sci. Technol.* **2019**, *53*, 3823–3831.
79. Jankovic, N. Z.; Plata, D. L. Engineered Nanomaterials in the Context of Global Element Cycles. *Environ. Sci.: Nano* **2019**, *6*, 2697–2711.
80. Adam, N.; Schmitt, C.; Galceran, J.; Companys, E.; Vakourov, A.; Wallace, R.; Knapen, D.; Blust, R. The Chronic Toxicity of ZnO Nanoparticles and $ZnCl_2$ to *Daphnia magna* and the Use of Different Methods to Assess Nanoparticle Aggregation and Dissolution. *Nanotoxicology* **2014**, *8*, 709–717.
81. Hunter, R. J. *Foundations of Colloid Science*, 2nd ed.; Oxford University Press: Oxford, 2001.
82. Tantra, R.; Bouwmeester, H.; Bolea, E.; Rey-Castro, C.; David, C. A.; Dogne, J. M.; Jarman, J.; Laborda, F.; Laloy, J.; Robinson, K. N.; Undas, A. K.; van der Zande, M.

Suitability of Analytical Methods to Measure Solubility for the Purpose of Nanoregulation. *Nanotoxicology* **2016**, *10*, 173–184.

83. David, C.; Galceran, J.; Rey-Castro, C.; Puy, J.; Companys, E.; Salvador, J.; Monné, J.; Wallace, R.; Vakourov, A. Dissolution Kinetics and Solubility of ZnO Nanoparticles followed by AGNES. *J. Phys. Chem. C* **2012**, *116*, 11758–11767.
84. Zhang, Y.; Muhammed, M. Critical Evaluation of Thermodynamics of Complex Formation of Metal Ions in Aqueous Solutions VI. Hydrolysis and Hydroxo-complexes of Zn^{2+} at 298.15 K. *Hydrometallurgy* **2001**, *60*, 215–236.
85. Domingos, R. F.; Simon, D. F.; Hauser, C.; Wilkinson, K. J. Bioaccumulation and Effects of CdTe/CdS Quantum Dots on *Chlamydomonas reinhardtii* — Nanoparticles or the Free Ions? *Environ. Sci. Technol.* **2011**, *45*, 7664–7669.
86. Domingos, R. F.; Franco, C.; Pinheiro, J. P. Stability of Core/shell Quantum Dots-role of pH and Small Organic Ligands. *Environ. Sci. Pollut. Res.* **2013**, *20*, 4872–4880.
87. Duval, J. F. L.; Farinha, J. P. S.; Pinheiro, J. P. Impact of Electrostatics on the Chemodynamics of Highly Charged Metal-Polymer Nanoparticle Complexes. *Langmuir* **2013**, *29*, 13821–13835.
88. Vale, G.; Franco, C.; Diniz, M. S.; dos Santos, M. M. C.; Domingos, R. F. Bioavailability of Cadmium and Biochemical Responses on the Freshwater Bivalve *Corbicula fluminea* — the role of TiO_2 nanoparticles. *Ecotoxicol. Environ. Saf.* **2014**, *109*, 161–168.
89. Vale, G.; Franco, C.; Brunnert, A. M.; dos Santos, M. M. C. Adsorption of Cadmium on Titanium Dioxide Nanoparticles in Freshwater Conditions — A Chemodynamic Study. *Electroanalysis* **2015**, *27*, 2439–2447.
90. Biver, T.; Friani, R.; Gattai, C.; Secco, F.; Tine, M. R.; Venturini, M. Mechanism of Indium(III) Exchange Between NTA and Transferrin: A Kinetic Approach. *J. Phys. Chem. B* **2008**, *112*, 12168–12173.
91. Pla-Vilanova, P.; Galceran, J.; Puy, J.; Companys, E.; Filella, M. Antimony Speciation in Aqueous Solution Followed with AGNES. *J. Electroanal. Chem.* **2019**, *849*, 113334.
92. Monteiro, A. C. S.; Companys, E.; Puy, J.; Galceran, J. How much analyte is removed from the electrode in Stripping ChronoPotentiometry? The depletion factor. *J. Electroanal. Chem.* **2024**, in press.
93. Krezel, A.; Wojcik, J.; Maciejczyk, M.; Bal, W. May GSH and L-His contribute to Intracellular Binding of Zinc? Thermodynamic and Solution Structural Study of a Ternary Complex. *Chem. Commun.* **2003**, 704–705.
94. Ferretti, L.; Elviri, L.; Pellinghelli, M. A.; Predieri, G.; Tegoni, M. Glutathione and N-acetylcysteinylglycine: Protonation and Zn^{2+} complexation. *J. Inorg. Biochem.* **2007**, *101*, 1442–1456.
95. Diaz-Cruz, M. S.; Díaz-Cruz, J. M.; Mendieta, J.; Tauler, R.; Esteban, M. Soft- and Hard-modeling Approaches for the Determination of Stability Constants of Metal-peptide Systems by Voltammetry. *Anal. Biochem.* **2000**, *279*, 189–201.
96. Harris, W. R.; Chen, Y.; Wein, K. Equilibrium-Constants for the Binding of Indium(III) to Human Serum Transferrin. *Inorg. Chem.* **1994**, *33*, 4991–4998.

https://doi.org/10.1142/9789811281099_0003

3 Metal–Ligand Speciation Using Electromigration Techniques

Jean Aupiais

Commissariat à l'Energie Atomique et aux Energies Alternatives,
Direction des Applications Militaires (CEA/DAM/DIF),
91297 Arpajon Cedex, France
Jean.AUPIAIS@CEA.FR

Table of Contents

I. List of Abbreviations

ε-AMC	ε-aminocaproic acid
CE	capillary electrophoresis
cIEF	capillary isoelectric focusing
Cit^{3-}	citrate
DFT	density functional theory
$DOTA^{4-}$	1,4,7,10-tetraazacyclododecane-1,4,7,10-tetraacetate
$DTPA^{5-}$	diethylenetriaminepentaacetate
$EDTA^{4-}$	ethylenediaminetetraacetate
EXAFS	extended X-ray absorption fine structure
HB	hydroxybutyrate
HIB	hydroxyisobutyrate
ICP-MS	inductively coupled plasma mass spectrometry
MES	2-(*N*-morpholino)ethanesulfonic acid
MOPS	3-morpholino-1-propanesulfonic acid
NMR	nuclear magnetic resonance
NTA^{3-}	nitrilotriacetate
pI	isoelectric point
TBACl	tetrabutylammonium chloride

T*f*	transferrine
TRIS	tris(hydroxymethyl)aminomethane
UV-vis	ultraviolet-visible spectroscopy

II. Principles of Electromigration

Electromigration techniques, which include capillary electrophoresis and isotachophoresis, are techniques that rely on the differential migration of charged species under the action of an electric field. All of these techniques obey the same laws, which are briefly described in this section. In an electrolyte, any charged ions that are present will migrate under the influence of an electric field E and with a velocity v that is proportional to this electric field (Eq. 1),

$$v = \mu E \ (\text{m s}^{-1}) \tag{1}$$

where μ is the electrophoretic mobility expressed in $\text{m}^2\ \text{V}^{-1}\ \text{s}^{-1}$.

Electrophoretic mobility, which is constant for a given ion and a given environment, is the resultant of the electrostatic forces attempting to move the charged ions and the frictional forces limiting their movement. At equilibrium, these opposing forces balance each other out, giving the following well-known relationship (Eq. 2),

$$\mu = \frac{ze}{6\pi\eta r} \tag{2}$$

where z is the charge of the ion, e is the elementary charge, η is the viscosity of the solution and r is the radius of the ion, which is assumed to be spherical (hence the 6π term).

This law is well defined at infinite dilution only and does not take into account the presence of any other ions in the vicinity of the central ion. Nonetheless, as will be seen below, the proportionality between μ and ze/r appears to be preserved at other ionic strengths; this means that when experiments are carried out using electrolytes with an identical composition, it becomes possible to obtain simple relationships among the electrophoretic mobilities of the species present in the solution. To take account of the ionic strength and explicitly define the effects of other parameters such as the temperature or permittivity of the solution, Eq. 3 is used. This equation, which was proposed by Fuoss and Onsager, makes it possible to quantify with good precision the physicochemical effects on electrophoretic mobility. It consists of two terms: a relaxation term (electrostatic correction) and an electrophoretic term (hydrodynamic correction). So, for an ionic strength I of less than 0.1 M, the Fuoss-Onsager equation given below can be used.[1]

$$\mu = \mu_0 - \left(\frac{e^3}{12\pi} \sqrt{\frac{\mathcal{N}_a}{(\varepsilon k_B T)^3}} \times z\mu_0 \sum_{n=0}^{\infty} C_n R^{(n)} + \frac{e^2}{6\pi\eta} \sqrt{\frac{\mathcal{N}_a}{\varepsilon k_B T}} |z| \right) \frac{\sqrt{I}}{1 + \frac{Ba}{\sqrt{2}} \sqrt{I}} \quad (3)$$

In Eq. 3, μ_0 is the intrinsic mobility of the ion at zero ionic strength, $\mathcal{N}_a$ is the Avogadro number, ε is the permittivity of the solution, k_B is the Boltzmann constant, $\sum_{n=0}^{\infty} C_n R^{(n)}$ is the contribution of all other ions to the electrostatic braking caused by the ionic atmosphere of the ion in question, I is the ionic strength and $Ba/\sqrt{2}$ is an adjustable parameter generally assumed to be equal to 1.5.[2]

The contribution of the other ions to the electrostatic braking, $\sum_{n=0}^{\infty} C_n R^{(n)}$, is calculated by determining the n components of the vector $R^{(n)}$ using the recursive formula of Eq. 4.

$$r_j^{(n)} = (2\mathbf{H} - 1)_{ji} r_i^{(n-1)}; r_j^{(0)} = r_j \quad (4)$$

The h_{ji} components of the matrix $\mathbf{H}$ are calculated by Eq. 5.

$$h_{ji} = \delta_{ji} \sum_i \phi_i \frac{\omega_i}{\omega_i + \omega_j} + \phi_i \frac{\omega_j}{\omega_i + \omega_j} \quad (5)$$

$\delta_{ji} = 1$ for $i = j$ and is equal to 0 in the other cases, $\omega_i = \frac{\lambda_i^0}{\lambda_i^0 + \lambda_j^0}$ and φ_i is the mole fraction. C_n is the nth member of a series calculated in the following manner:

$$C_n = \frac{\prod_{i=0}^{n-1} (p - i)}{n!} \times \frac{-\sqrt{2}}{2} \quad (6)$$

with $C_0 = 1/2(2 - \sqrt{2})$ and $p = 1/2$. In practice, one stops at C_5, because the level of precision has already reached four decimal places.

This type of calculation may seem difficult, but by following the step-by-step explanations given in Fuoss and Onsager's original article,[1] one can fairly easily set up an Excel™-type spreadsheet for $n = 3$ ions that are present simultaneously. Beyond this number, it is advisable to use software such as PeakMaster,[3] which was developed by Gas and co-workers.[4-6]

It might seem strange that the movement of an ion in solution that is subjected to an electric field could provide information about the environment surrounding the ion. Indeed, the immediately observable feature is only a speed (and therefore an electrophoretic mobility), which is unlikely to provide any particular information about the type of ion or its surrounding environment. On its own, a speed is uninformative, but when a speed varies due to a modification of the environment surrounding the ion, this is a situation that, to the contrary, does provide information. Extracting this hidden information relies heavily on hypotheses, which should be relatively strong and unambiguous. The whole challenge in carrying out speciation studies using electromigration techniques therefore consists in ensuring that the variations of speed measured are due only to the object being studied and not to external, independent causes. Such causes are unfortunately numerous, which means that the experimental conditions must be controlled in an almost metrological fashion.

Based on a physical (*i.e.*, an electric field) rather than on a chemical principle (*i.e.*, a chemical reaction), electromigration techniques can be considered to be global probes, as opposed to spectroscopic techniques, which are local probes. Under certain conditions that will be detailed later in this chapter, these probes can give access to intrinsic information pertaining to the molecule or the ion such as its charge, its degree of oxidation, the intramolecular distance within a metal-ligand complex, and the number and types of complexes formed.

III. Methods

A. Capillary Zone Electrophoresis

Capillary zone electrophoresis is the method most frequently used for speciation studies. This popularity stems from the method's great ease of use, its impressive versatility and the option of coupling several types of detectors (ultraviolet-visible spectroscopy (UV-vis), fluorescence, mass spectrometry, or conductimetry). The separations are performed inside a fused silica capillary that is coated with a polyimide polymer to make it highly flexible. Because the capillaries that are used are made of fused silica, one must take account of the ionization of the silanol groups on the inner surface of the capillary. This ionization appears around pH 2 and reaches a maximum at pH 8. When an electric field is applied to the terminals of the capillary, a flow referred to the electro-osmotic flow appears, contributing to the speed of the ions (see Figure 1). The speed of the electro-osmotic flow is defined by Eq. 7,

$$v_{\text{eof}} = -\frac{\varepsilon_r \varepsilon_0 \zeta}{\eta} E = \mu_{\text{eof}} E \tag{7}$$

where μ_{eof} is the mobility of the electro-osmotic flow, ε_r and η are respectively the relative dielectric constant and the viscosity of the solvent, ε_0 is

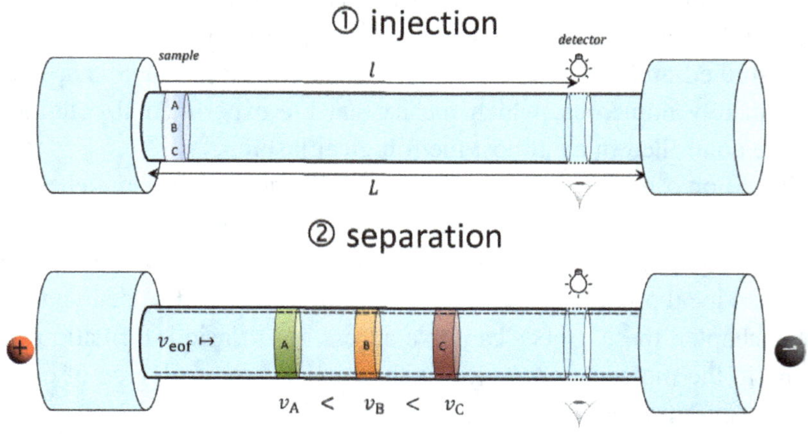

Figure 1. Principle of capillary zone electrophoresis with spectrophotometric detection. ① The sample containing the analytes A, B, and C is injected at the inlet of a capillary, of length L, that is filled with a conductive electrolyte. ② After the electric field is applied (here, the example pertains to cations) and the electrolyte is set into motion by electro-osmosis, the ions to be analyzed are separated due to their different speeds ($v_A < v_B < v_C$), which are proportional to their electrophoretic mobilities ($\mu_A < \mu_B < \mu_C$) and increased by the speed of the electro-osmotic flow (v_{eof}). They are detected at a distance l in a zone in which the protective polyimide sheath surrounding the capillary has been removed to allow light to pass into the detector.

the vacuum permittivity ($\varepsilon_0 = 8.85 \times 10^{-12} C^2N^{-1}m^{-2}$) and ζ is the zeta potential.

The zeta potential is determined by the number of charges on the surface of the capillary and is defined by Eq. 8,

$$\zeta = \frac{\kappa^{-1}\sigma}{\varepsilon_r \varepsilon_0} \tag{8}$$

where κ^{-1} is the thickness of the double layer (also known as the Debye length) and σ is the charge density per unit length. The thickness of the double layer is given by Eq. 9,

$$\kappa^{-1} = \sqrt{\frac{\varepsilon_r \varepsilon_0 RT}{2c_E z^2 F^2}} \tag{9}$$

where R is the perfect gas constant, T is the absolute temperature, F is the Faraday constant, and z and c_E are respectively the charge and the concentration of the electrolyte.

Therefore, the speed of an ion is the sum of its own speed and that of the electro-osmotic flow (Eq. 10).

$$v = \mu E + \mu_{eof} E = (\mu + \mu_{eof})E \tag{10}$$

As the electro-osmotic flow is always directed towards the cathode, it accelerates the cations and decelerates the anions. The electro-osmotic flow can be reversed by coating the internal surface of the capillary with a cationic surfactant, either dynamically or semi-permanently (for example with the cations cetyltrimethylammonium,[7] didodecyldimethylammonium,[8] *etc.*) or permanently (for example with trimethylammoniumstyrene).[9] This strategy is often applied to accelerate the anions and shorten the analysis time. Still, for speciation studies involving labile species, the electro-osmotic flow is an essential parameter that must be systematically determined for the electrophoretic mobilities to be quantified. From an experimental viewpoint, the quantification of the electrophoretic mobilities demands that the migration time of the species be determined as well. The following relationship is used (Eq. 11),

$$\mu = \frac{lL}{V}\left(\frac{1}{t} - \frac{1}{t_{\text{eof}}}\right) \tag{11}$$

where L is the total length of the capillary, l is the length of the capillary up to the detector ($l < L$) (see Figure 1), t is the migration time of the species and t_{eof} is the migration time of the electro-osmotic flow. When a mass spectrometer is used, $l = L$, because the detector is situated after the capillary.

Species detection by capillary zone electrophoresis depends on whether the complexes involved are stable or labile (see Speciation section). With stable complexes, the parameter that must be taken into account is the surface area (or the corrected surface area). With labile complexes, the migration time is what provides information about the speciation. In this latter case, the value of the electro-osmotic flow must be known so that the electrophoretic mobilities can be corrected accordingly.

As for the quantification of stable species, it depends on the type of detector used. In the case of photometric detectors (UV-vis, fluorescence), the corrected surface area expressed according to Eq. 12 is used,

$$S_{\text{c}} = \frac{S}{t} \tag{12}$$

where S is the surface area of the peak and t is the migration time.

Detection is carried out by sampling the variation of the signal in a specific zone of the capillary (Figure 1) with a frequency that generally ranges between 2 and 32 Hz. Indeed, an electrophoregram is generally represented by correlating each measuring point of the detector with a time (namely a migration time). Yet capillary electrophoresis separation is based on differences of speed in a given electric field. What the detector actually "sees" are ions crossing it at variable speed (see Figure 2). Because the sampling step is constant, the surface area calculated for ions that are present in the same concentration and that respond in an identical fashion will be greater for slower ions and smaller for faster ions. To compensate for this effect and reduce the degree of uncertainty, it is advisable to use Eq. 12 rather than the surface area alone.

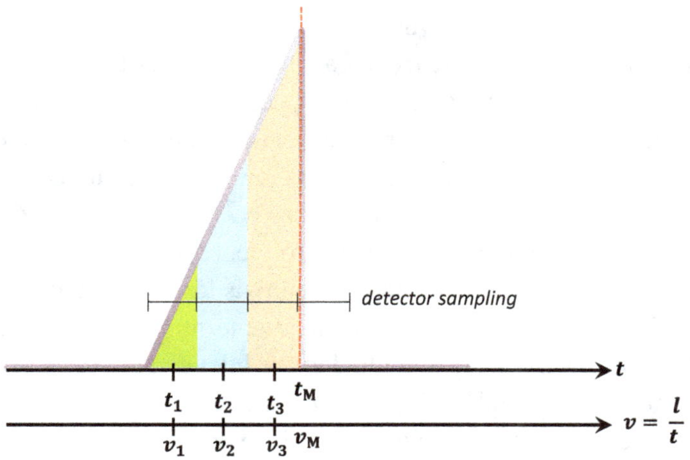

Figure 2. Principle of spectrophotometric detection in capillary electrophoresis. The detector takes samples with a constant step as the ions crossing it do so at different speeds. The principle is the same for a peak corresponding to only one type of ion; the speed of the ions at the tailing edge of the peak (here v_1) is slower than that of the ions close to the apex (v_3). The resulting distortion of the net surface area of the peak is corrected by dividing it by the migration time of the entire peak at the maximal concentration (t_M).

When capillary electrophoresis is coupled with mass spectrometry, ion detectors are used. The ions are detected once and once only. In this configuration, no correction is necessary, and so it is the surface area of the peak S that must be used rather than S_c.

B. Isotachophoresis

Isotachophoresis is a method exclusively based on the separation of charged molecules according to their electrophoretic mobility. The main detectors used are the same as those used for capillary electrophoresis, namely UV-vis and conductivity. In the case of isotachophoresis, another detector that is specific to this technique has often been used: the potential gradient detector, which detects the variation of the electric field between the zone boundaries. Isotachophoresis uses two electrolytes, rather than only one as in the case of capillary zone electrophoresis. All of the ions move at the same speed, with the most mobile ones in the lead and the slowest ones bringing up the rear. In practice, a capillary tube is filled with an electrolyte composed of an ion (either positive or negative, but of the same charge as the ions to be separated) whose mobility is greater than that of all of the ions to be separated. This ion is referred to as the leader ion, and its speed is v_L. After the sample has been injected at the capillary inlet, the capillary is immersed in an electrolyte, referred to as the terminal electrolyte, comprising an ion whose mobility is lower than those of all of the ions to be separated. This ion is called the terminal ion, and its speed is v_T. When the electric field is applied to the inlet and outlet of the capillary, a weak electric field will be set in place in the leader electrolyte, whereas a strong electric field will be established in the terminal electrolyte. A weak field slows the ions down, but a strong field speeds them up. When equilibrium is reached, a number of stable zones appear, with each of them being composed of one of the ions A, B, C, *etc.*, to be separated, and with each of these ions being characterized by an individual mobility μ_A, μ_B, μ_c, *etc.* (see Figure 3). On a practical level, it may be noted that:

1. The leader and terminal ions will move at the same speed and will flank the species of interest, as follows (Eq. 13):

$$v_L = \mu_L \times E_L = \mu_T \times E_T = v_T \tag{13}$$

Because of these identical speeds and because μ_L and μ_T are set to be constant, the electric fields E_L and E_T compensate for the differences between the electrophoretic mobilities.

2. The ions to be analyzed (A, B, C, *etc.*) will gradually separate until the following equality (Eq. 14) is satisfied.

$$v_L = v_A = v_B = v_C = \cdots = v_T \tag{14}$$

As the same causes produce the same effects, the electric fields that are associated with each migration band of the ions A, B, C, *etc.*, will compensate for the electrophoretic mobilities μ_A, μ_B, μ_C, *etc.* Between two zones, there exists an electric field gradient that depends on the

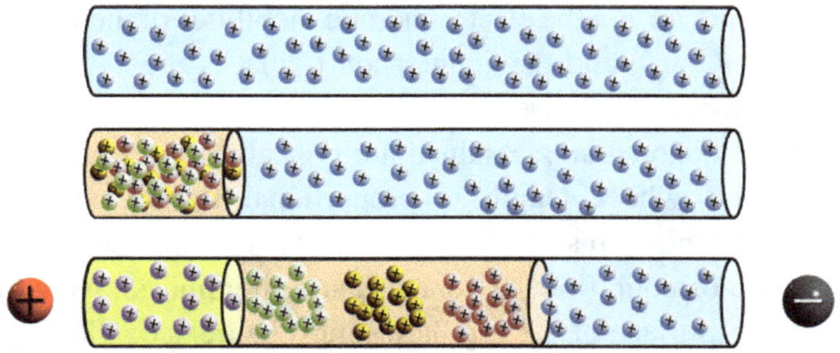

Figure 3. Principle of isotachophoresis. The figure illustrates the case in which cationic species are separated. The principle is identical for the analysis of anions, except that the polarity must be reversed; (uppermost image) the capillary is filled with a leader electrolyte whose leader cation has the highest mobility of all of the cations, (middle) the sample containing the species is injected at the inlet (the anode side); (bottom) The anode end of the capillary is immersed in a terminal electrolyte whose terminal cation has the lowest mobility of all of the cations. When an electric field is applied, stable zones are formed and are set into motion at the same speed. Each zone contains only cations with the same electrophoretic mobility.

difference of mobilities with regard to the neighboring ions. An ion that exits its zone, entering a field zone that is either stronger or weaker, is inevitably brought back into the zone from which it originated. This normalization of the speeds leads to a valuable characteristic of isotachophoresis: the migration bands are also focused. This technique is often used to concentrate trace elements (an enrichment factor of 200 is possible with this technique).[10]

3. The current intensity is the same everywhere in the capillary. On a practical level, if one uses solutions that are highly concentrated in leader and terminal ions (thereby promoting a high current intensity), then the concentrations of ions A, B, C, *etc.*, will increase in their respective zones. The ratio of the concentration of the ions of interest c_i to that of the leader electrolyte c_{L} can be calculated from the Kohlrausch regulation function[11] as follows (Eq. 15),

$$\frac{c_i}{c_{\mathrm{L}}} = \frac{\mu_i\left(\mu_L + \mu_{\bar{\mathrm{C}}}\right)}{\mu_{\mathrm{L}}\left(\mu_i + \mu_{\bar{\mathrm{C}}}\right)} \tag{15}$$

where μ_i, μ_L, and $\mu_{\bar{C}}$ are the electrophoretic mobilities of the ion of interest i, the leader ion L and the counter-ion $\bar{C}$ that is present both in the leader and terminal electrolytes.

The isotachophoregrams obtained are typical of this technique: they display plateaus whose length is proportional to the concentration (Figure 4). Isotachophoresis, which is a closed system, usually requires that the electro-osmotic flow be nonexistent or minimized, which means that the internal surface of the capillaries must be treated. Otherwise, the speed of the electro-osmotic flow would change continuously as the terminal electrolyte moves forward (and therefore as the leader electrolyte disappears), which would disturb the stability of the zones. It should be noted, however, that isotachophoresis with electro-osmotic flow (an open system) has been successfully tested, but requires a judicious selection of electrolytes.[12] Isotachophoresis with electro-osmotic flow is a technique with strong potential, because it allows for coupling with other detectors

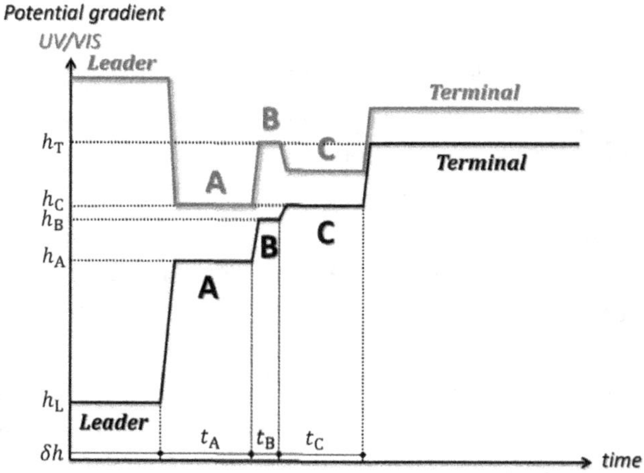

Figure 4. Isotachophoregrams obtained using UV-vis detection and potential gradient detection. t_i are the durations (equivalent to a length in the capillary) of the plateaus of the leader, A, B, C, and terminal ions; h_i are the respective step heights for the ions in question (with a potential gradient detector, because of an asymmetric potential resulting from the contamination of the electrode, the values of the step heights must be corrected by a value of $+\delta h$ obtained using an internal standard).[14] In isotachophoresis, lengths of the plateaus, not the step heights, are proportional to the analyte concentrations. Adapted from Ref. 14.

such as mass spectrometers. Unfortunately, even to this day, previous studies have dealt only with the quantitative analysis of organic substances, and never with speciation.[13] For further information, the reader is invited to refer to a recent contribution written by a specialist of this technique, T. Hirokawa.[14]

C. Capillary Isoelectric Focusing

Isoelectric focusing is a technique that "self-preconcentrates" the ampholytes. Ampholytes such as proteins, amino acids, or water can behave simultaneously like an acid and like a base. Therefore, depending on the pH, these species can be either positively or negatively charged. Between the two states, there is consequently a pH value for which the overall charge of the ampholyte is zero. Thus, by applying an electric field in a capillary subjected to a pH gradient, the ampholytes can be made to migrate into a zone (namely a pH) where the overall charge will be zero. Indeed, an ampholyte located in the acidic zone will be positively charged and will migrate towards the cathode. The same ampholyte located in a basic zone of the capillary will be negatively charged and will migrate towards the anode. The focal point corresponds to an electrophoretic mobility of zero, and the pH value found is an experimental measurement of its isoelectric point (referred to as *pI*) or point of zero charge. The difficulty resides in creating a stable pH gradient within the capillary, which in practice means creating static zones, each of them characterized by a different pH. This problem was overcome in the 1970's through the use of carrier ampholytes.[15] Like isotachophoresis, capillary isoelectric focusing (cIEF) requires the use of coated capillaries to suppress the electroosmotic flow. The goal of obtaining a regular pH gradient creates the need for a large number of ampholytes with slightly different *pI* values covering a pH range suitable for the sample. In addition, they must present an electrical conductivity in the vicinity of their *pI* in order to maintain the conductivity of the electrolyte. The gels employed for cIEF must contain at least about 30 ampholytes per unit of pH. Due to the very strong electric fields used in cIEF (for better focusing), commercial ampholytes can contain more than 1000 ampholytes per unit of pH.[15]

Separation by cIEF is a three-step process (Figure 5). Sometimes a spacer is added to the mixture (in step ①) to stabilize the pH gradient on the cathode and/or anode side. The spacer creates a buffer zone between the compartments and the capillary, with the width of this zone corresponding to the amount of spacer added. The most common spacers are acetic acid for the acidic side, and *N,N,N',N'*-tetramethylethylenediamine for the basic side.[15]

The focusing is controlled by the current. When the current becomes stable, the focusing is complete, because this means that the ampholytes are no longer moving.[16] Detection is carried out by "pushing" the electrolyte towards the detector, which is very frequently a UV-vis detector due to its suitability for use with proteins. This mobilization can be performed through the application of pressure while keeping the electric field unchanged in order to maintain the same pH gradient and therefore the same focusing. However, mobilization through pressure produces a laminar flow that is detrimental to the resolution. It is therefore preferable to move the solution at a rate slower than 1 cm min^{-1} to prevent a significant loss of resolution. Another, more ingenious, method is often used instead, because it enables the resolution to be preserved. This method consists in adding a salt (such as NaCl) to one of the electrode solutions. For example, if H_3PO_4 and NaOH are respectively used as the anolyte and the catholyte, the addition of NaCl into either compartment will provoke within the capillary the movement of the species towards the reservoir where the salt has been added. For example, if NaCl is added to NaOH, the Na^+ and OH^- ions will be introduced into the pH gradients of the cathode and the anode, respectively. The pH gradient will therefore be increased due to the presence of the OH^- ions, which will in turn lead the ampholytes to become negatively charged and consequently to migrate towards the anode. If NaCl is added to H_3PO_4, the H^+ ions introduced into the pH gradient will lower its value; the ampholytes become positively charged which lead to an electrophoresis on the cathode side. The important thing to remember with this mobilization technique is that the H^+ or OH^- ions must be replaced by other cations or anions, respectively.

Because the separation of the ampholytes is based on the isoelectric points, the addition of *pI* markers to the solution enables the quantification of the isoelectric points. In general, three markers are sufficient to obtain

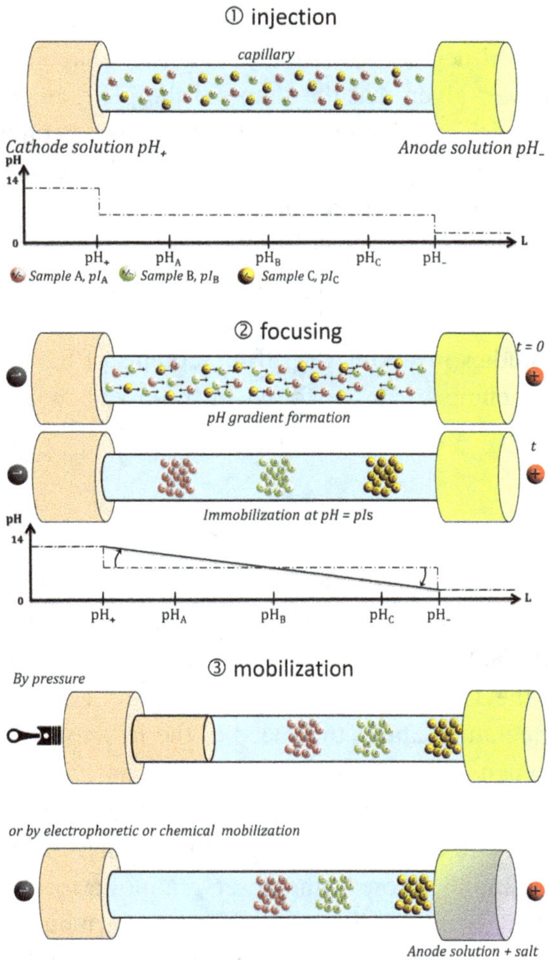

Figure 5. Principle of capillary isoelectric focusing. Step ①: The capillary is entirely filled with an electrolyte containing a support gel, the sample to be analyzed (here there are three ampholytes A, B and C), carrier ampholytes (suited to the desired pH range) and possibly anodic and/or cathodic spacers (whose role is to stabilize the pH gradient should it become necessary); there is no need for a preference regarding the positions of the catholyte and anolyte or their pH; the present example involves NaOH at the cathode (basic catholyte, pH_+) and H_3PO_4 at the anode (acidic anolyte, pH_-). Step ②: On the application of the electric field (whose direction depends on the position of the acidic and basic solutions at the electrodes), a pH gradient is established, and the ampholytes (both the sample and carrier ampholytes) start to migrate towards a point that satisfies the equality pH = *pI*. Once the current is stabilized, the immobilization is finished. Step ③: The mobilization can be carried out either by pressure or by electrophoretic mobilization (also referred to as chemical mobilization). A salt, for example NaCl, is then added to one of the compartments. The mobilization begins, moving in the direction of the compartment to which the salt was added.

a linear pH-distance relationship and thereby quantify the unknown *pI* values with a good degree of accuracy. One point should be monitored vigilantly, however: given that the shape of the pH gradient depends on the composition of the carrier ampholytes, the focusing conditions and the detection system, linearity over a wide range of pH often goes undemonstrated. It is more useful to bracket an unknown value by two markers than to carry out a calibration.[15]

Capillary isoelectric focusing coupled with inductively coupled plasma mass spectrometry (ICP-MS) is a particularly valuable technique for speciation studies involving the affine proteins of toxic metals. More details on this technique can be found in the overview provided by K. Shimura.[15]

IV. Prerequisites

The charge, the metal-ligand distance and the order of the complexes formed can be accessed only if all of the parameters influencing electrophoretic mobility are controlled. The main factors that absolutely must be controlled are the temperature and the ionic strength.[17] A one-degree increase in temperature causes the speed of the ions to increase by about 2%. This phenomenon depends on the type of supporting electrolyte used. Also, a small variation in the total concentration of ions comprising the supporting electrolyte can lead to temperature changes from one experiment to the next and therefore to the observation of a speed variation that does not result from the studied species, but from modifications of the physico-chemical properties of the supporting electrolyte, such as the viscosity and the permittivity. The second factor, ionic strength, also affects these parameters. While this effect is less significant than that caused by temperature fluctuations, it can nonetheless lead one to believe that the local environment of the ion in question has been modified, when in fact what causes its electrophoretic mobility to vary is only an overall effect of the electrolyte in its entirety. An example of the effect of ionic strength is given in Figure 6. The objective of the study was to examine the hydrolysis reaction of the anionic species $[Th(DTPA)]^-$ ($DTPA^{5-}$ = diethylenetriaminepentaacetate) by determining the variation of the

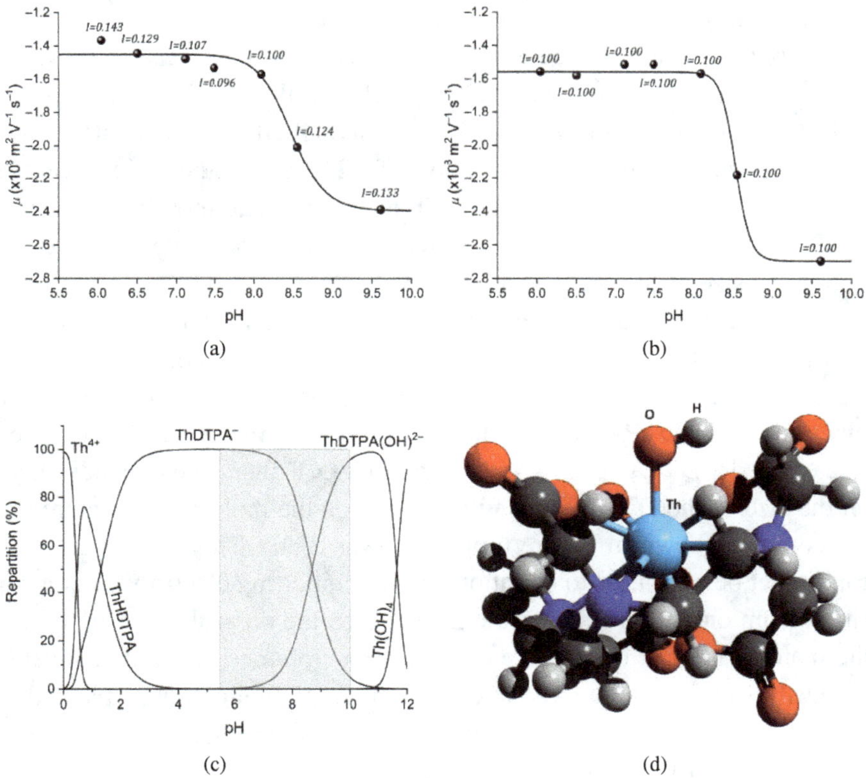

Figure 6. Variation of the electrophoretic mobility of the $Th^{4+}/DTPA^{5-}$ species as a function of pH, $C_{Th} = 10^{-8}$ M, $C_{DTPA} = 10^{-2}$ M. According to the speciation diagram (c), at pH values between 5.5 and 10 (shaded rectangle), the variation in mobility corresponds to the transition from the $[Th(DTPA)]^-$ species to the $[Th(DTPA)(OH)]^{2-}$ complex. (a) Raw uncorrected values obtained at the ionic strength mentioned alongside each point. The reaction constant determined by fitting the experimental points (solid line) is $\log K = 8.47 \pm 0.14$. (b) Values following correction to a single ionic strength of $I = 0.1$ M using the Fuoss-Onsager equation (Eq. 3).[1] The adjusted constant becomes $\log K = 8.53 \pm 0.04$. (d) Snapshot of the structure of the $[Th(DTPA)(OH)]^{2-}$ complex calculated using quantum molecular dynamics.

electrophoretic mobility between pH 6 and pH 10 for the following reaction equilibrium (Eq. 16).[18]

$$\left[\mathrm{Th}(\mathrm{DTPA})\right]^{-} + \mathrm{OH}^{-} \rightleftharpoons \left[\mathrm{Th}(\mathrm{DTPA})(\mathrm{OH})\right]^{2-} \qquad (16)$$

According to the speciation diagram shown on the lower left of Figure 6, the average charge carried by the Th-DTPA species should vary from –1.002 at pH 6 (*i.e.*, 99.8% $[Th(DTPA)]^-$ + 0.2% $[Th(DTPA)(OH)]^{2-}$) to –1.024 at pH 7 (*i.e.*, 97.6% $[Th(DTPA)]^-$ + 2.4% $[Th(DTPA)(OH)]^{2-}$), which comes to a variation of –2.2%, all other things being equal. In practice, an 8.1% decrease in electrophoretic mobility is measured experimentally. This variation could have been attributed to a variation in the size of the species (a hydroxyl group is added to the complex), but theoretical quantum molecular dynamics calculations have shown that the distance between the hydroxyl group and the central atom ($r = 2.15$ Å) is shorter than the distances between the oxygen atoms of the carboxylic functions of the $DTPA^{5-}$ and of this same central atom ($r = 2.35$ Å).[18] The size of the Th-DTPA species with or without OH^- therefore depends only on the size of the $DTPA^{5-}$ chelating agent (see the molecular structure of the complex shown in Figure 6 on the lower right). This 8.1% variation can in fact be attributed to variations in the supporting electrolyte composition from one experiment to the next. Given the difficulties involved in the manipulation of radioactive elements, the volumes manipulated were reduced to about one hundred microliters in total, inducing variable additions of several microliters to several tens of microliters of the DTPA solution and for the purposes of pH adjustment.

In practice, the ionic strength varied from 0.096 M to 0.143 M for a theoretical target of 0.1 M (see Figure 6, upper left). Correcting electrophoretic mobility to a single ionic strength is performed by iteration, either by programming Eq. 3 into an Excel™-type spreadsheet or by using PeakMaster,[3] a freeware program developed by Gas and co-workers. The methodology is as follows: based on the experimental measurements, the type of supporting electrolyte (its concentration, viscosity and dielectric permittivity)[17] and the ionic strength value, the absolute mobility μ_0 is determined by iterating the Fuoss-Onsager equation. Once this value is known, all one needs to do is to adjust the ionic strength to the desired value, with all other parameters being set. This correction, which is illustrated in Figure 6 (upper right), allows for better reproducibility and better final accuracy.

Finally, there is another relatively subtle effect that modifies electrophoretic mobility (and therefore the speed) by increasing the relaxation term (electrostatic correction). In the Fuoss-Onsager equation (Eq. 3), the term $\sum_{n=0}^{\infty} C_n R^{(n)}$ represents the contribution of all of the other ions to electrostatic braking. This term applies in the same way to all species present in the solution. During the separation of species using a technique such as capillary electrophoresis, however, the species that are co-migrating within the same migration band will have an influence on each other, while those that are completely separated will be influenced only by the ions making up the supporting electrolyte.[19] Fortunately, this effect becomes significant only when the concentrations of co-migrating species exceed 10^{-3} M.[19] It is negligible when capillary electrophoresis is coupled with ICP-MS (CE–ICP-MS), because the concentrations being studied are always much lower than the millimolar level due to the very high sensitivity of the apparatus. On the other hand, if other, less sensitive detectors are used (for example, UV-vis), care must be taken to ensure that the concentration of co-migrating species does not exceed 10^{-3} M. This point requires special vigilance, especially if these co-migrating species are invisible to the detector being used. So, for instance, with UV-vis detection, any non-absorbing species that is present in the migration band of the studied species could modify its electrophoretic mobility and therefore lead to an incorrect conclusion about its speciation.

If these precautions are taken, electromigration techniques can supply valuable information about the speciation of metals such as the charge, the degree of oxidation, the relative sizes of the complexes, and so on.

V. Charge

It is impossible to quantify the effective charge of a species, because according to Eq. 2, electrophoretic mobility depends on both the charge and the inverse of the Stokes radius. To extract an effective charge value, one must "force" the system to respond to this variable by selecting a physical phenomenon that unequivocally responds to a charge variation, regardless of the size of the species in question. An ideal probe therefore interacts with ionic species without interfering with them on a chemical level. Such probes are common in analytical chemistry; they generally rely on UV-vis or fluorescence spectroscopy. Because the energy provided by these photons is not sufficient to induce parasitic chemical reactions, it therefore represents an ideal inert probe. That said, in order to respond to the charge variation in a quantitative manner, the probe must be non-specific. Consequently, this method relies on the use of chromophores that present absorption or fluorescence properties at a given wavelength. In practice, a chromophore of the same charge sign as the ions to be analyzed is added to the supporting electrolyte. In the case of absorption for example, the change in absorption ΔA linked to the expulsion of part of the chromophore by the species of interest is described by Eq.17,

$$\Delta A = (R\varepsilon_c - \varepsilon_i)lC_i \tag{17}$$

where R is the displacement ratio (*i.e.*, the number of chromophore molecules displaced per molecule or ion studied), ε_c is the molar extinction coefficient of the chromophore at the operating wavelength, ε_i is the molar extinction coefficient of the molecule or ion studied, l is the optical path, and C_i is the concentration of the molecule or ion studied. The variable that responds quantitatively to the charge is the displacement ratio R. In cases where the species of interest does not absorb at the absorption wavelength of the chromophore, the change of absorbance becomes directly proportional to R if the concentration of the C_i species remains unchanged. Since the response depends on the molar extinction coefficient of the chromophore, a calibration procedure must be carried out for each set of operating conditions (see Table 1). For example, Figure 7 shows a linear variation of the corrected surface area of the electrophoretic peaks (see the section on Capillary Zone Electrophoresis) of a number of cations, at the same concentration and as a function of their charge ($r^2 = 0.996$). It is therefore possible to evaluate the real effective charge of any species that is analyzed under the same conditions and at the same concentration as the species used to perform the calibration.[20]

VI. Degree of Oxidation

Electromigration techniques, especially capillary electrophoresis, have been very widely used to detect and quantify the oxidation states of metallic cations ever since the first commercial apparatuses appeared on the market.[23-26] Capillary electrophoresis is a non-intrusive method that does not alter oxidation states, provided that the supporting electrolyte is not

Table 1. Surface area of the peaks for cations of charge z = 1, 2, and 3, at a set concentration of 0.1 mM, for a selection of chromophores (r is the linear correlation coefficient).

Chromophore	$z = +1$	$z = +2$	$z = +3$	Slope	r
1-naphthylamine	9 ± 2	17 ± 4	40 ± 4	15	0.961
UV Cat 1*	24 ± 5	50 ± 6	69 ± 6	23	0.995
Imidazole	17 ± 3	35 ± 2	65 ± 7	24	0.988
Phenyltrimethylammonium	11 ± 2	19 ± 6	62 ± 6	25	0.926
Ephedrine	17 ± 5	36 ± 10	72 ± 7	28	0.988
Pyridine	38 ± 9	74 ± 8	109 ± 11	35	0.999
UV Cat 2*	64 ± 15	128 ± 19	177 ± 56	56	0.999

* Products patented and commercialized by Waters.[21,22]

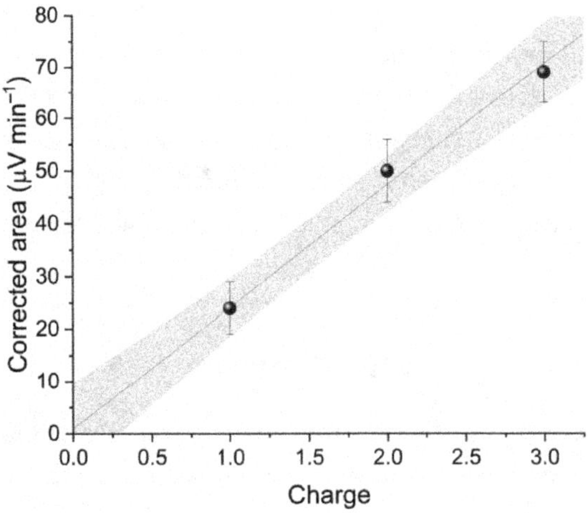

Figure 7. Relationship between the effective charge and the corrected surface area for a number of cations of identical concentration (C_i = 0.1 mM). Average corrected surface area of n ions with z = +1 (n = 7: Li^+, NH_4^+, Na^+, K^+, Rb^+, Cs^+, Ag^+), z = +2 (n = 10: Be^{2+}, Mg^{2+}, Ca^{2+}, Mn^{2+}, Ni^{2+}, Co^{2+}, Zn^{2+}, Sr^{2+}, Ba^{2+}, Pb^{2+}), z = +3 (n = 16: Cr^{3+}, Y^{3+}, La^{3+}, Ce^{3+}, Pr^{3+}, Nd^{3+}, Sm^{3+}, Eu^{3+}, Gd^{3+}, Tb^{3+}, Dy^{3+}, Ho^{3+}, Er^{3+}, Tm^{3+}, Yb^{3+}, Lu^{3+}). Operating conditions: chromophore [UVCat 1] = 5 mM (λ = 185 nm), V = 20 kV, i = 5–6 μA, pH = 4.4 (adjustment with HNO_3). Adapted from Ref. 20.

"redox active." The literature often features reviews dealing with the determination of the redox states of various metallic cations.[23,24,26-29] Often these reviews involve toxic metals and non-metals such as arsenic, lead, mercury, nickel, cobalt, chromium, copper, zinc, the actinides and lanthanides, or essential metals such as iron or selenium.[24,30] It is not uncommon for these techniques to be used for "exotic" elements or for those for which the characterization of oxidation states is especially difficult when using other techniques. For example, the astatide (At^-) anion, whose degree of oxidation is –1, was first experimentally identified by means of electromigration by Guo and co-workers in 2018, and its electrophoretic mobility has been measured experimentally under acidic reducing conditions.[31] Recent applications are also found in the nuclear industry (speciation of rhenium in oxidation states IV and VII,[32] speciation of the actinides in oxidation states III, IV, V and VI),[33,34] and in the battery manufacturing industry (speciation of manganese in oxidation states II and III).[35]

Very few studies deal with the determination of oxidation states using isotachophoresis. Hirokawa *et al.* showed that Fe^{II} could be separated from Fe^{III} from their chelates with phenanthroline by using as the leader electrolyte a 20 mM ammonia solution (pH 4.8 with acetic acid) and as the terminal electrolyte a 20 mM hydrochloric acid solution (pH 3.6 with β-alanine).[25]

VII. Structure

It may seem surprising that a separation technique could provide any information about the structure of a complex. In fact, according to Eq. 2, electrophoretic mobility is inversely proportional to the size of the ion, with all other things being equal. This dependency is, however, masked by other influential parameters such as charge and viscosity. It is nonetheless possible to extract structural information using one of the two methods detailed below. The first method requires the intramolecular distances to be externally calibrated, either by experimental measurements (such as extended X-ray absorption fine structure, or EXAFS) or by computational chemistry (density functional theory, or DFT). The second method is completely experimental but requires that the considered complexes carry full effective charges.

A. Determination of Intramolecular Metal–Ligand Distances by External Calibration

For species of identical structure that differ only according to their central cation, the size variations caused by any differences in the lengths of the chemical bonds between the ligand and the cation ought to modify the electrophoretic mobility value. However, given the expected smallness of these variations (roughly a few tenths of an angstrom), it is wise to consider whether they would have a detectable effect on electrophoretic mobility. In fact, the discovery of this access to intramolecular distances in a metal-ligand complex occurred through pure chance. It originated from an experimental observation made during a study focusing on the complexation of actinides in the +IV oxidation state by the polyaminocarboxylate ligand $DTPA^{5-}$. The study dealt with the determination of the complexation constants of the Th^{4+}, U^{4+}, Np^{4+}, and Pu^{4+} series through CE–ICP-MS.[18] In order to optimize the machine time, all cations were simultaneously injected to perform a single separation process rather than four individual separations. One of the experiments carried out with a metal/ligand concentration ratio of 10^{-5} was sufficiently intriguing for the authors to opt for a more detailed investigation (see Figure 8).[19]

As Figure 8 (right) shows, electrophoretic mobility (which can be evaluated based on migration time) decreases as the size of the ion

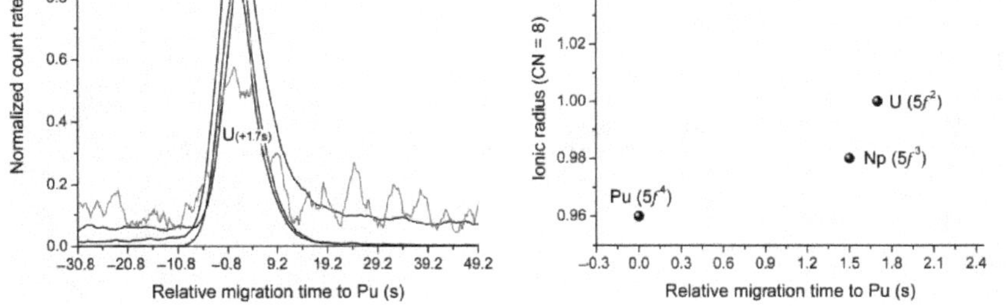

Figure 8. (left) Relative migration time (with Pu as the reference) of the $[An^{IV}(DTPA)]^-$ species ($An^{4+} = Th^{4+}, U^{4+}, Np^{4+}, Pu^{4+}$) obtained through CE–ICP-MS. Operating conditions: $C_{DTPA} = 10^{-2}$ M, $C_{An} = 10^{-7}$ M, supporting electrolyte: $NaClO_4$, $I = 0.1$ M, pH = 6.045, $V = 10$ kV, $L = 59.6$ cm, $T = 298$ K, injection at 0.5 psi for 3 s. (right) Variation of the migration time (with Pu as the reference) as a function of ionic radius for a coordination number of 8. Adapted from Ref. 19.

increases. The variation observed nonetheless does not follow a simple rule, because it depends on the electronic configuration of the ions. It is therefore appropriate as a first step to measure or calculate the intramolecular cation–ligand distances, and then, as a second step, to correlate the values obtained with the electrophoretic mobilities. The differences in migration times are so small and so dependent on the operating conditions that the best way to establish a relationship between electrophoretic mobility and intramolecular distance is to measure these differences with respect to a given species. Here, thorium was selected as the reference mobility (Figure 9). To correlate a variation in mobility (y axis) with an intramolecular distance (x axis), intramolecular distances must be obtained by other means. Here, EXAFS measurements were carried out on pure complexes of three actinides (Th, Np and Pu) and confirmed by DFT calculations (see Table 2). The linear relationship that emerges in Figure 9 (left) makes it possible to calculate, by interpolation, the intramolecular distance between uranium and the five binding oxygen atoms of the five carboxylate functions of the $DTPA^{5-}$ in the $[An^{IV}(DTPA)]^-$ complex. Let us consider as an example the $[U(DTPA)]^-$ species, which is unstable when irradiated and whose structure could not be determined by EXAFS. Two electrophoretic mobility values for the $[U(DTPA)]^-$ complex

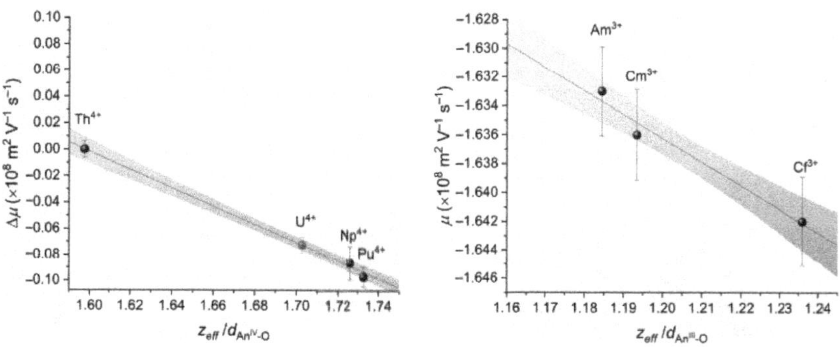

Figure 9. (left) Electrophoretic mobility difference $\Delta\mu$ with respect to $[Th(DTPA)]^-$ for the $[An^{IV}(DTPA)]^-$ species obtained under four different pH conditions ($An^{4+} = Th^{4+}$, U^{4+}, Np^{4+}, Pu^{4+} in grey shading) as a function of the ratio of the effective charge z_{eff} to the average intramolecular distance $d_{An\cdots O}$ of the five carboxylate functions of $DTPA^{5-}$. (right) The same relationship obtained using a single experiment with the $[An^{III}(DTPA)]^{2-}$ system ($An^{3+} = Am^{3+}$, Cm^{3+}, Cf^{3+}).

Table 2. Average distances between the An^{4+}/An^{3+} cation and the five oxygen atoms of the carboxylate functions of $DTPA^{5-}$, as well as the effective charge (z_{eff}) of An^{4+}/An^{3+}.

An^{4+}	$d_{An^{IV}\cdots O}/Å$ (EXAFS) [19]	z_{eff} [36]	An^{3+}	$d_{An^{III}\cdots O}/Å$ (DFT) [37]	z_{eff} [36]
Th^{4+}	2.39	+3.82	Pu^{3+}	—	+2.888
U^{4+}	—	+3.88	Am^{3+}	2.44	+2.890
Np^{4+}	2.30	+3.97	Cm^{3+}	2.43	+2.900
Pu^{4+}	2.29	+3.97	Cf^{3+}	2.39	+2.954

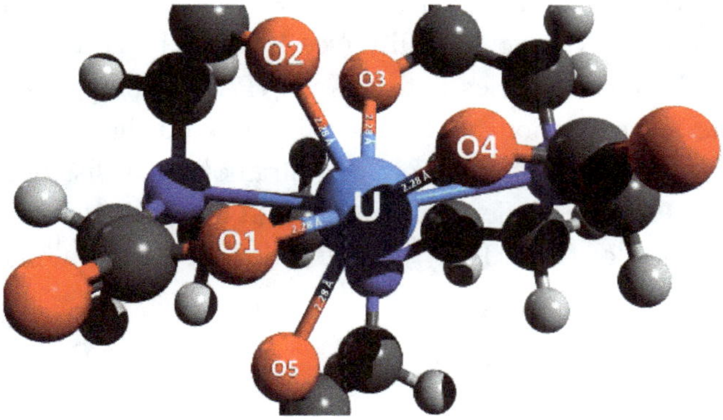

Figure 10. Diagram of the $[U(DTPA)]^{-}$ complex (U–O intramolecular distance obtained using capillary electrophoresis).

were found using CE–ICP-MS. So, for an average experimental value $\Delta\mu_{Th-U} = 0.0725 \times 10^{-8}$ m^2 V^{-1} s^{-1}, the extrapolated value $z_{eff}/d_{An\cdots O}$ = 1.702 is obtained from Figure 9. Due to the hardness of actinides (the driving force is mainly electrostatic), the effective charge z_{eff} is used instead of the true charge z because the higher-energy electrons do not experience the full nuclear charge of the nucleus owing to the shielding effect of the inner-shell electrons. Since z_{eff} = 3.88 for U^{4+},[36] the distance $d_{U\ldots O}$ = 2.28 Å is found (a diagram of the complex is given in Figure 10). This value perfectly agrees with other evaluations based on thermodynamics (2.28 Å), *ab initio* calculations (2.30 Å) and even EXAFS spectroscopy (2.33 Å following extrapolation from the values of $[Pu(DTPA)]^{-}$,

$[Th(DTPA)]^-$ and $[Np(DTPA)]^-$).[19] Remarkably, a precision within a few hundredths of an angstrom is reached (± 0.05). It must be emphasised, however, that this type of measurement requires an almost metrological degree of control over the operating conditions (temperature, ionic strength, *etc.*). Indeed, in this study, the variations in electrophoretic mobility among the complexes come to only about 10^{-10} $m^2\ V^{-1}\ s^{-1}$, representing only a few percent of the measured electrophoretic mobility values.

Likewise, similar experiments were carried out with the $[An^{III}(DTPA)]^{2-}$ system. At the maximal DTPA concentration of 10^{-2} M, three of the actinides considered exhibited nearly identical mobilities, much like the observation made for the tetravalent actinides. Only Pu^{3+} presented a significantly lower mobility (see Figure 11). At this concentration, the predominant species is the $[An(DTPA)]^{2-}$ complex (more than 99.9%). The linear relationship that is also observed should make it possible to determine, by extrapolation, the intramolecular distance $d_{Pu^{III}\cdots O}$ in the $[Pu(DTPA)]^{2-}$ complex. In fact, up until now, all attempts to elucidate the structure of $[Pu(DTPA)]^{2-}$ have failed, because Pu^{III} is unstable

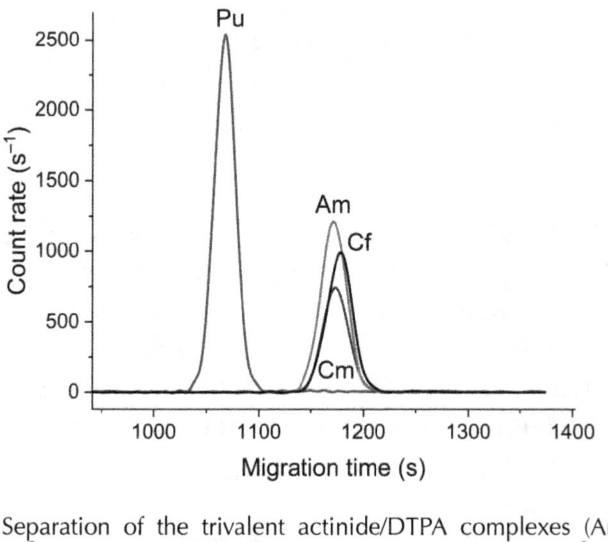

Figure 11. Separation of the trivalent actinide/DTPA complexes ($An^{3+} = Pu^{3+}$, Am^{3+}, Cm^{3+}, Cf^{3+}) by CE–ICP-MS. Operating conditions: $C_{DTPA} = 10^{-2}$ M, $C_{An} = 5 \times 10^{-7}$ M, supporting electrolyte: NaCl, $I = 0.1$ M, pH = 2.02, $V = 10$ kV, pressure: 0.8 psi, $L = 59.8$ cm, $T = 298$ K, injection at 1 psi for 4 s.

and is easily oxidized by irradiation during X-ray absorption experiments. As shown in Figure 11, however, the peak of the $[Pu(DTPA)]^{2-}$ complex is not found at the position that was expected based on the behavior of the other actinides. The explanation lies in the weaker complexation constant of Pu^{3+} ($\log \beta_{Pu} = 20.58$, $\log \beta_{Am,Cm} = 23.1$, $\log \beta_{Cf} = 23.7$),[37,38] which does not allow the limiting complex to be formed. One can conclude from this example that this $d_{Pu^{III}\cdots O}$ distance is perfectly accessible when electromigration techniques are used. The solution is fairly "simple": it suffices to increase the ligand strength by working at a higher pH so that the limiting complex is reached for all of the actinides.

B. Determination of the Coordination Modes of Labile Metal–Ligand Complexes

The ability to discern small variations in electrophoretic mobility can be used to elucidate the coordination modes of a series of complexes with the same central cation but different numbers of ligands of a single type. As an example, let us consider the complexation of pentavalent (AnO_2^+) or hexavalent (AnO_2^{2+}) actinide cations with common anions such as carbonate. The reaction equilibrium is written as follows (Eq. 18).

$$AnO_2^{x+} + nCO_3^{2-} \rightleftharpoons \left[AnO_2\left(CO_3\right)_n\right]^{(x-2n)+} \tag{18}$$

Structural studies by EXAFS have shown that the coordination mode of the carbonate ligand with hexavalent plutonium always seems to be bidentate.[39,40] For the pentavalent actinides, the EXAFS spectra are difficult to interpret, but the authors agree that the coordination mode is also bidentate.[41] That said, the most recent thermodynamic data show an anomaly for the 1:3 complexes of pentavalent neptunium and plutonium, suggesting that a change has occurred in the coordination mode of at least one carbonate ligand.[42] Several mobility values are available at two different ionic strengths ($I = 0.37$ M and $I = 0.2$ M),[42,43] which allows Eq. 2 to be tested. The results are shown in Figure 12. It is remarkable to note that, despite its simplicity, this equation remains valid provided that all experiments to be compared are performed using the same experimental conditions (same ionic strength, viscosity, temperature, supporting electrolyte,

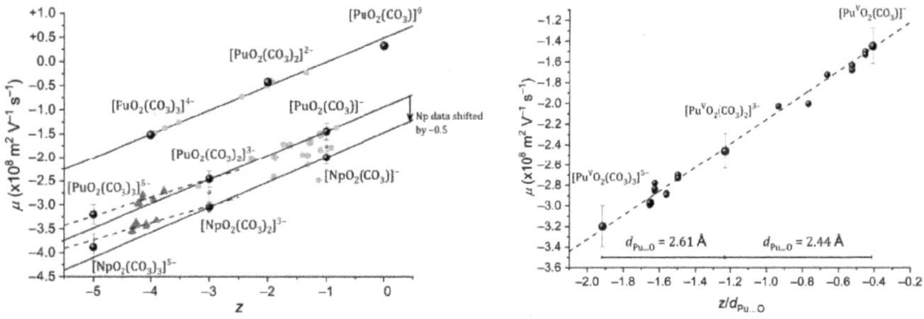

Figure 12. (left) Variation of the electrophoretic mobilities of Pu^{VI}, Pu^{V}, and Np^{V} carbonato complexes as a function of their charge. Black dots: mobility of the pure complexes obtained after mathematical processing (see Speciation section), grey dots: average mobility corresponding to mixtures of two adjacent complexes in varying proportions. The resulting average charge is calculated using the complexation constants associated with the two complexes in question. Data relative to Np^{V} have been shifted by 0.5×10^{-8} to improve visibility. (right) The same variations of the mobilities of the Pu^{V} carbonato complexes, corrected using the ratio $z/d_{Pu\cdots O}$ ($d_{Pu\cdots O}$ is obtained using quantum molecular dynamics). Adapted from Ref. 42.

etc.). Assuming that all of the carbonates adopt a bidentate coordination mode, the size of the complex remains the same regardless of the number of ligands ($n = 1$, 2, or 3), because the ligands are merely exchanged with molecules of water (see Figure 13). This results in a mobility of the complexes that depends only on their charge. This is precisely what is observed in Figure 12 (left) for the charged Pu^{VI} carbonate complexes and the first two Pu^{V} complexes, for which the mobility varies proportionally to charge alone.

One can nonetheless note the presence of a deviation for the 1:3 complexes of Pu and Np, which suggests that the coordination mode has changed. By now using the results of the quantum molecular dynamics calculations determining the average distances between the binding oxygen atoms and the central cation, it can be seen that one of the carbonate ligands adopts a monodentate coordination mode (Figure 13) with a distance $d_{An...O} = 2.61$ Å, whereas the bidentate carbonates are all at a constant distance $d_{An...O}$ 2.44 Å. Thus, by expressing the electrophoretic mobility as a function of $z/d_{An...O}$, a linear relationship is obtained (Figure 12, right). Provided the separations are performed using the same

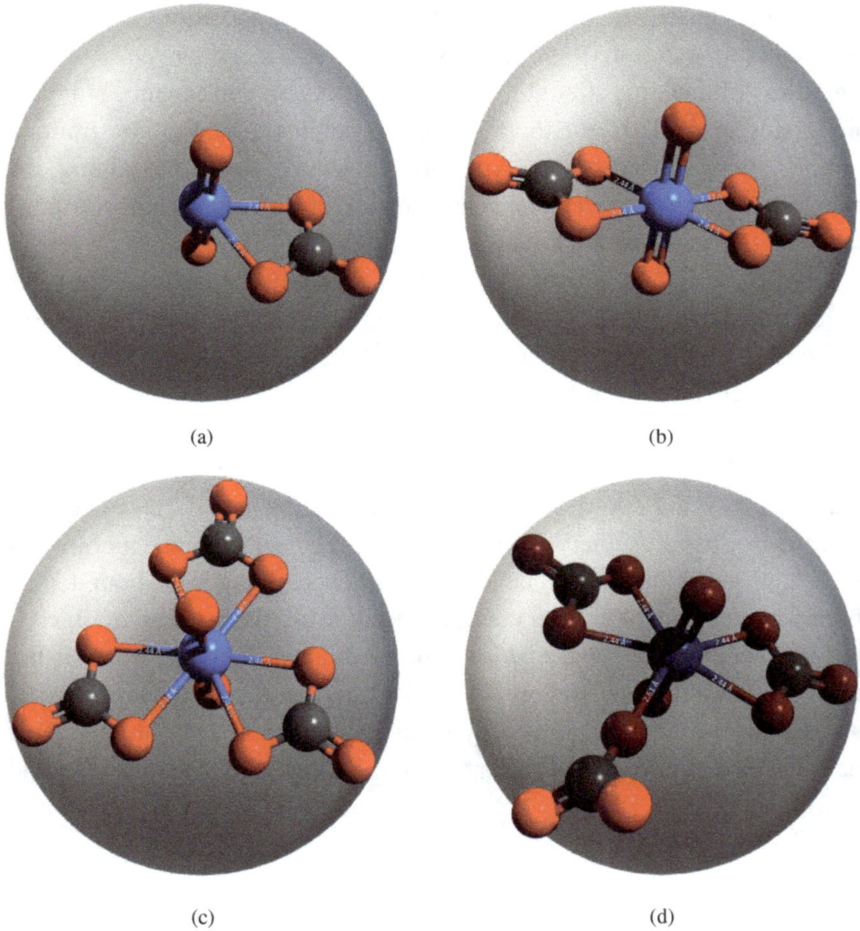

(a) (b) (c) (d)

Figure 13. Hydration model of the actinyl-carbonate complexes. The coordination sphere is completed by water molecules (not shown) that are present in the spherical volume (shaded in grey). Structures of (a) the 1:1 complexes ($[An^{V}O_2(CO_3)]^-$, $[An^{VI}O_2(CO_3)]$) and (b) the 1:2 complexes ($[An^{V}O_2(CO_3)_2]^{3-}$, $[An^{VI}O_2(CO_3)_2]^{2-}$) in bidentate coordination mode. Structure of the 1:3 complexes: (c) for Pu^{VI}, all carbonate ligands are bidentate; (d) for Pu^{V}, DFT calculations show that two carbonate ligands are bidentate and one is monodentate. In the latter case, the greater length of the monodentate bond leads to a greater overall volume for the complex and therefore a lower mobility than in a configuration in which all of the carbonate ligands are bidentate (case of Pu^{VI}). The increase in volume is due to the two oxygen atoms (in pale red) that are "exiting" the volume occupied by the complex in a totally bidentate configuration.

supporting electrolyte (*i.e.*, an identical ionic composition), this result shows that the relation of proportionality between μ and z/r remains valid regardless of the ionic strength. This has significant consequences for data processing, especially for thermodynamic studies aimed at determining complexation constants (see Speciation section). By constraining the mobility values in this way, the determination of the equilibrium constants becomes more precise.

In case of deviation from the ze/r proportionality, it becomes possible to determine the increase in size of the complex, and therefore the longest of the distances between the carbonate and the central cation. Since Eq.2 has been verified, it can be applied to the $[PuO_2(CO_3)_3]^{5-}$ complex. One on hand, one can find its mobility μ_1, considering that the carbonate ligands adopt a completely bidentate coordination mode. Quantum molecular dynamics and EXAFS experiments give a distance of 2.44 Å for this coordination mode. Otherwise, one can obtain μ_2, which is the mobility of this same complex when one of the carbonate ligands is bound in a monodentate fashion to the central cation, with the two other ligands being bidentate. The value of μ_1 is estimated using the linear relationship in Figure 12 (left), working on the assumption that the coordination mode is exclusively bidentate, while μ_2 is an experimental measurement that makes no assumption about the nature of the complex. Hence, Eqs. 19 and 20 follow.

$$\mu_1 = \frac{-5}{6\pi\eta r_1} \tag{19}$$

$$\mu_2 = \frac{-5}{6\pi\eta r_2} \tag{20}$$

The ratio of these mobilities is:

$$\frac{\mu_1}{\mu_2} = \frac{r_2}{r_1} \tag{21}$$

With $\mu_1 = -3.476 \times 10^{-8}\ m^2\ V^{-1}\ s^{-1}$, $\mu_2 = -3.323 \times 10^{-8}\ m^2\ V^{-1}\ s^{-1}$, and $r_1 = 2.44$ Å, the numerical application of Eq. 21 (Figure 12) provides

$r_2 = 2.62 \pm 0.07$ Å, which is in very good agreement with the radius calculated by DFT simulations ($r = 2.61 \pm 0.05$ Å).

VIII. Speciation

Electromigration techniques, and capillary electrophoresis in particular, are often used to determine equilibrium constants. These studies are based on the variation of electrophoretic mobility as a function of ligand concentration. These studies can be classified into two types according to whether stable or labile complexes are identified. In the case of stable complexes, there are as many electrophoretic peaks as there are species. For labile complexes, an equilibrium is established among all of the species in the migration band of the element interacting with the ligand.[44]

A. Case of Stable Complexes

1. *Capillary electrophoresis*

Some ligands, especially organic ligands, show particularly high affinities for metallic cations. The complexation constants in such situations can be well above 10^{10} and even reach values as high as 10^{50}. In practice, it is difficult to set a boundary between the two cases (strong and weak complexes), but one can often note that constants of about 10^{20} lead, under appropriate conditions, to the observation of as many peaks as there are species. Interpretation is therefore often simplified, because the number of species present simultaneously is never very high. For strong complexes, the key parameter is the surface area of the peak. Let us consider the following example of an exchange reaction between two ligands L_1 and L_2 (Eq. 22).

$$ML_1 + L_2 \rightleftharpoons ML_2 + L_1 \tag{22}$$

The corresponding equilibrium constant is expressed by Eq. 23.

$$K = \frac{[ML_2][L_1]}{[ML_1][L_2]} \tag{23}$$

If the detector response is the same for ML_1 and ML_2, the special case $[ML_1] = [ML_2]$ corresponds to the situation in which the surface area of the peak of the L_1 complex is equal to the surface area of the peak of the L_2 complex. Thus, the complexation constant can be easily calculated in this case and depends only on the ratio of the concentrations of the free ligands $[L_1]$ and $[L_2]$. This method is mainly used in studies focusing on the complexation of metals by proteins. All one needs to do is to gain access to one of the complexation constants, using any method, in order to determine the other. It can indeed be easily demonstrated that in this situation, there is a simple relationship (Eq. 24) between the constants,

$$K = \frac{K_2}{K_1} \tag{24}$$

where K_1 and K_2 are the formation constants of complexes ML_1 and ML_2, respectively.

An example is given in Figure 14. It shows the variations in surface area of the species $[Pu(NTA)_2]^{2-}$ (NTA^{3-} = nitrilotriacetate) and Pu*Tf* (the

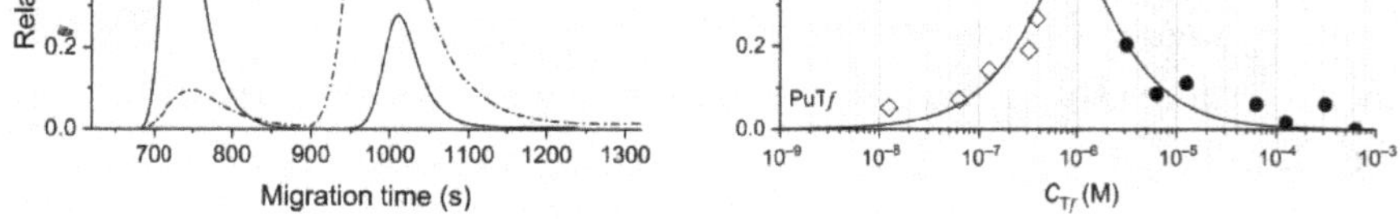

Figure 14. (left) Separation by capillary electrophoresis of the $[Pu(NTA)_2]^{2-}$(L_1 = NTA^{3-}) species and the Pu*Tf* transferrin complex (L_2 = *Tf*) for two ligand concentration ratios: $[L_1]/[L_2] = 20$ (solid line) and $[L_1]/[L_2] = 0.01$ (dotted line). Separation conditions: $T = 298$ K, $V = 5$ kV, Beckman Coulter N-CHO™ capillary, $L = 65$ cm, $\varnothing_{internal} = 50\ \mu$m, pH = 6.0, supporting electrolyte: NaCl/2-(*N*-morpholino)ethanesulfonic acid, $I = 0.1$ M, $C_{Pu} = 10^{-10}$ M, $[L_1] = 10^{-6}$ M. (right) Variation of the ratio of the surface areas of the two peaks as a function of transferrin concentration for a constant NTA concentration of 10^{-6} M. Adapted from Ref. 45.

Pu-transferrin complex) for two ligand concentration ratios, respectively 20/1 and 1/100. The equilibria considered in these experiences are given by Eqs. 25 and 26.

$$Pu^{4+} + 2NTA^{3-} \rightleftharpoons \left[Pu(NTA)_2\right]^{2-} \qquad \beta_2 = \frac{\left[\left[Pu(NTA)_2\right]^{2-}\right]}{\left[Pu^{4+}\right]\left[NTA^{3-}\right]^2} \tag{25}$$

$$Pu^{4+} + Tf \rightleftharpoons PuTf \qquad \beta^* = \frac{[PuTf]}{\left[Pu^{4+}\right][Tf]} \tag{26}$$

The formation constants associated with these two ligands are very large: $\log \beta_2 = 37.90$ with NTA^{3-} and $\log \beta^* = 24.76$ for T*f*, respectively.[45] The formation constant of PuT*f* was obtained at the inflection point of Figure 14 (right) given the formation constant of $[Pu(NTA)_2]^{2-}$.[18]

2. *Isotachophoresis*

As with capillary electrophoresis, strong complexes must be studied in an acidic environment in order to measure a conditional constant that is small enough to show both the metal and the complex. In this situation, however, the isotachophoretic zone containing the anionic metal-ligand complex gradually liberates the metallic cation due to a decomposition phenomenon affecting the complex that shortens the length of the zone. It has, however, been demonstrated that on one hand, the shortening of the zone depends on the value of the complexation constant, and that on the other hand, this zone is isotachophoretically stable.[46] This zone shortening can therefore be referred to for quantitative purposes.[47] Table 3 below summarizes the conditions used in two studies dealing with this type of complex.

B. Case of Labile Complexes

1. *Capillary zone electrophoresis*

For labile complexes, the equilibration time among the various species is much shorter than the separation time. In practice, this results in a single

Table 3. List of studies dealing with the determination of stability constants (labile complexes) by isotachophoresis.

Cation	Ligand	Leader electrolyte	Terminal electrolyte	Ref.
Lanthanides	EDTA	20 mM HCl (pH 3.1 by triglycine; pH 3.6 by β-alanine; pH 4.4 by ε-AMC; pH 6.0 by histidine)	20 mM NH_3 (pH 4.8 by acetic acid; pH 4.8 by caproic acid; pH 6.0 by MOPS)	47
Ho^{3+}, Y^{3+}	DTPA, DOTA	20 mM HCl (pH 3.1 by glycine)	20 mM NH_3 (pH 4.8 by acetic acid)	48

$EDTA^{4-}$ = ethylenediaminetetraacetate; ε-AMC = ε-aminocaproic acid; $DTPA^{5-}$ = diethylenetriaminepentaacetate; $DOTA^{4-}$ = 1,4,7,10-tetraazacyclododecane-1,4,7,10-tetraacetate.

migration band containing all of the species of interest in equilibrium with each other. In this case, the key parameter is not the surface area, because all of the species contribute to it, but rather the electrophoretic mobility. The overall electrophoretic mobility $\bar{\mu}$ is thus a linear combination of all electrophoretic mobilities μ_i associated to the species i weighted by their molar fractions x_i (Eq. 27).

$$\bar{\mu} = \sum_i x_i \mu_i \tag{27}$$

The variation of electrophoretic mobility in a migration band containing the metal and the ligand can thus be described using the following relationship (Eq. 28),

$$\bar{\mu} = \frac{\mu_M + \sum_i \beta_i \mu_i [L]^i}{1 + \sum_i \beta_i [L]^i} \tag{28}$$

where μ_M is the electrophoretic mobility of the uncomplexed cation, μ_i is the electrophoretic mobility of the complex i, β_i is the global stability constant for the formation of complex i and [L] is the free ligand concentration. Eq. 28 is the outcome of successive equilibria between a metallic cation M and an anionic ligand L. As an example, let us consider a system in which the cation binds a maximum of three ligands. This gives three successive equilibria (Eqs. 29–31).

$$M + L \rightleftharpoons ML, K_1 = \frac{[ML]}{[M][L]} \tag{29}$$

$$ML + L \rightleftharpoons ML_2, K_2 = \frac{[ML_2]}{[ML][L]} \tag{30}$$

$$ML_2 + L \rightleftharpoons ML_3, K_3 = \frac{[ML_3]}{[ML_2][L]} \tag{31}$$

It should be recalled that β_i in Eq. 28 is related to the successive constants K_i by the relationship $\beta_i = \Pi_i K_i$.

An electrophoretic mobility is associated to each complex. Therefore, for each equilibrium, the total mobility variation has a sigmoidal shape, with the plateaus corresponding to the mobilities of the two complexes (see Figure 15). Summing these curves leads to a complex curve that is generally sigmoidal in shape as well, but with small deviations that indicate the presence of other complexes. Mathematical processing therefore

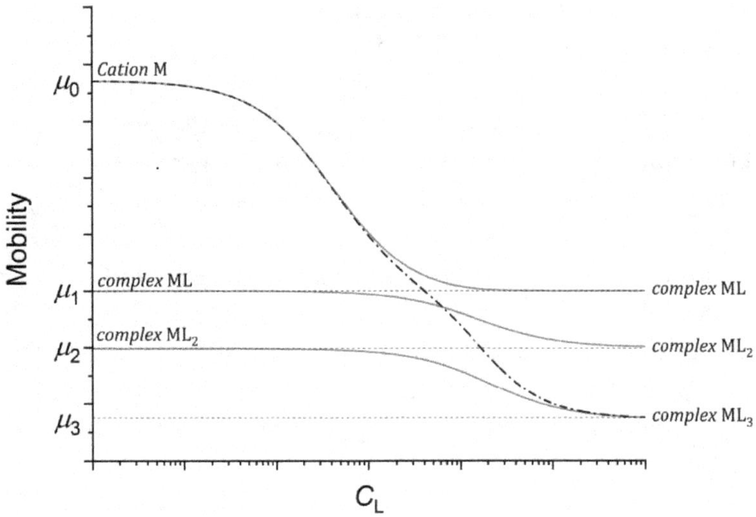

Figure 15. Individual variations of the electrophoretic mobilities for the system $ML_i + L \rightleftharpoons ML_{i+1}$ (i = 0–3). The black dotted curve is the sum of the three grey curves; it is consistent with experimental observations.

becomes the critical aspect in determining and quantifying the concentrations of the complexes that are present. In the case provided here as an example, the overlapping of the complexes demands the adjustment of seven parameters in Eq. 28 (four electrophoretic mobilities and three complexation constants). This level of difficulty makes it necessary to develop a strategy that aims not only to reduce the number of free variables where possible, but also to seek independent relationships among variables in view of limiting the number of possible solutions.

Eq. 28 must be used with caution. This is due to the interdependency of the parameters μ and β; the product $\beta \times \mu$ creates a situation in which the overestimation of one variable leads to an underestimation of the other. It must, however, be noted that the magnitude of the electrophoretic mobilities is generally limited to a few 10^{-8} m^2 V^{-1} s^{-1}, even though the possible stability constant values are spread over several orders of magnitude. Surprisingly, though, the results obtained are quite robust. As a matter of fact, uncertainties of 30% for electrophoretic mobility have been observed to lead to uncertainties of only a few tenths of a logarithmic unit for the β constant. Nonetheless, to reduce as far as possible the uncertainties associated with the mobilities, a few simple rules should be observed (see Table 4), while considerable precision can be obtained by applying the empirical relationships in Table 5. It has indeed been experimentally observed that the ratios of the electrophoretic mobilities of metallic complexes with the same ligand to their charge difference were constant. The coefficient, however, is different depending on whether the complexes are positively or negatively charged. The slope is about 1.7×10^{-8} for cationic complexes but only 0.5×10^{-8} for anionic complexes. For the moment, no firm explanation can be given for this difference in behavior between cations and anions, but perhaps a lesser degree of hydration of the anions could be one factor. As for cationic complexes, it should be noted that the slope is almost identical to that which would have been obtained by calculating theoretical values using Eq. 32 proposed by Anderko and Lencka.[49]

$$\mu_{ML_n} = \frac{|z_{ML_n}|}{\sqrt[3]{\left(\frac{|z_M|}{\mu_M}\right)^3 + n\left(\frac{|z_L|}{\mu_L}\right)^3}} \tag{32}$$

Table 4. Rules and parametric constraints to be applied for the mathematical adjustment of the function $\overline{\mu} = f(\beta_i, \mu_i)$ (Eq. 28) for complexation reactions involving a metallic cation M^{z+} and a ligand L^{z-}.

Case*	Rule	Comments
$[M(L)_n]^0$ $z^+ + nz^- = 0$	$\mu \approx 0$	In the case of ICPMS detection, because the effective mobility is calculated with respect to a neutral marker, the difference is not always strictly equal to 0 and can reach a few 10^{-9} mobility units ($m^2\ V^{-1}\ s^{-1}$).
$[M(L)_n]^{x+}$ $x = z^+ + nz^- > 0$	$\mu > 0$	
$[M(L)_n]^{x-}$ $x = z^+ + nz^- < 0$	$\mu < 0$	
Comparison of two complexes $[M(L)_{n_1}]^x$ and $[M(L)_{n_2}]^y$ $x < y$	$\mu_{[M(L)_{n_1}]^x} < \mu_{[M(L)_{n_2}]^y}$	The more highly charged complex is always faster.
Relationship between the complexes $[M(L)_{n_1}]^x$ and $[M(L)_{n_2}]^y$	$\alpha = \dfrac{\Delta\mu}{\Delta z} = \dfrac{\left\lvert \mu_{[M(L)_{n_2}]^y} - \mu_{[M(L)_{n_1}]^x} \right\rvert}{\lvert y - x \rvert}$ $\alpha_{cation} = 1.7$ $\alpha_{anion} = 0.5$	There is a relationship of proportionality between the mobilities of complexes with the same ligand (see Table 5). α is expressed in mobility units, *i.e.*, $10^{-8}\ m^2\ V^{-1}\ s^{-1}$.
	$\mu_{[M(L)_{n_2}]^y} \neq \dfrac{y}{x} \times \mu_{[M(L)_{n_1}]^x}$	It is incorrect to postulate that the relationships between complexes depend only on their charge ratios.

*convention used: $z^- < 0$

In contrast, by applying the same methodology, which operates on the assumption that the sign of the charge carried by the complex is of no importance, Eq. 32 does not replicate the experimental values for anionic complexes, as shown in Table 5. In fact, the equation was put forth solely to estimate the mobility of cationic complexes (see the original article by

Table 5. Experimental and theoretical values of the coefficient α for various systems. The theoretical α_{th} values were determined by using the electrophoretic mobilities of all complexes calculated by Eq. 32 and the electrophoretic mobilities for the cations and ligands taken at zero ionic strength.

Chemical equilibrium	Medium	$\alpha_{exp.} = \frac{\Delta\mu}{\Delta z}$ × 10^8/m^2 V^{-1} s^{-1}	α_{th}[49]	Ref.
$PuO_2^+ + nCO_3^{2-} \rightleftharpoons [PuO_2(CO_3)_n]^{(2n-1)-}$	0.37 M NaCl	0.51 ± 0.04	2.09	42
$NpO_2^+ + nCO_3^{2-} \rightleftharpoons [PuO_2(CO_3)_n]^{(2n-1)-}$	0.37 M NaCl	0.52 ± 0.05	2.09	42
$PuO_2^{2+} + nCO_3^{2-} \rightleftharpoons [PuO_2(CO_3)_n]^{(2n-2)-}$	0.20 M $NaClO_4$	0.50 ± 0.05	2.06	42
$UO_2^{2+} + nC_2O_4^{2-} \rightleftharpoons [UO_2(C_2O_4)_n]^{(2n-2)-}$	0.1 M NH_4Cl	0.49 ± 0.15	2.21	51
$Th^{4+} + nAc^- \rightleftharpoons [Th(Ac)_n]^{(n-4)-}$ $(n = 4,5)$	0.3 M Acetic acid	0.50	1.63	52
$La^{3+} + nC_2O_4^{2-} \rightleftharpoons [La(C_2O_4)_n]^{(2n-3)-}$	0.1 M NH_4Cl	0.94*	1.89	51
$Y^{3+} + nLac^- \rightleftharpoons [Y(Lac)_n]^{(3-n)+}$	0.015 M Benzylammonium	1.73 ± 0.04	2.02	32
$UO_2^{2+} + nAc^- \rightleftharpoons [UO_2(Ac)_n]^{(2-n)+}$	0.3 M Acetic acid	1.75 ± 0.12	1.70	52
$Th^{4+} + nAc^- \rightleftharpoons [Th(Ac)_n]^{(4-n)+}$	0.3 M Acetic acid	1.56 ± 0.03	1.69	52
$Am^{3+} + nAc^- \rightleftharpoons [Am(Ac)_n]^{(3-n)+}$	0.3 M Acetic acid	1.74 ± 0.21	1.74	52

*Calibration with only two points (n = 2 and 3). The mobility of the 1:2 complex ($\mu = -0.28 \times 10^{-8}$ m^2 V^{-1} s^{-1}) seems too low for a complex with a charge of −1.

Anderko and Lencka).[49] In contrast, anions seem to follow another, fully linear, rule. To conclude, it would not be appropriate to estimate the mobility of a complex based on the neighboring complexes by considering the charge to be the only parameter of influence (see last rule in

Table 4). On one hand, the application of Eq. 32 to cations demonstrates that no such rule can exist. On the other hand, for anions, Chandler observed through experiments that the ratios of the equivalent limiting conductivities of several diacids (between the HA^- and A^{2-} forms) were not equal to the ratio of their charges (0.5), but were in fact closer to 0.6.[50] This shows that even for such simple cases as that of diacids, it would be incorrect to postulate that there exists a relationship of proportionality that depends only on charge ratios.

For example, an application of the rules set out in Table 4 is illustrated in Figure 16 for the $UO_2^{2+}/C_2O_4^{2-}$ system. Three complex formation reactions (n = 1, 2, 3) are considered (Eq. 33).

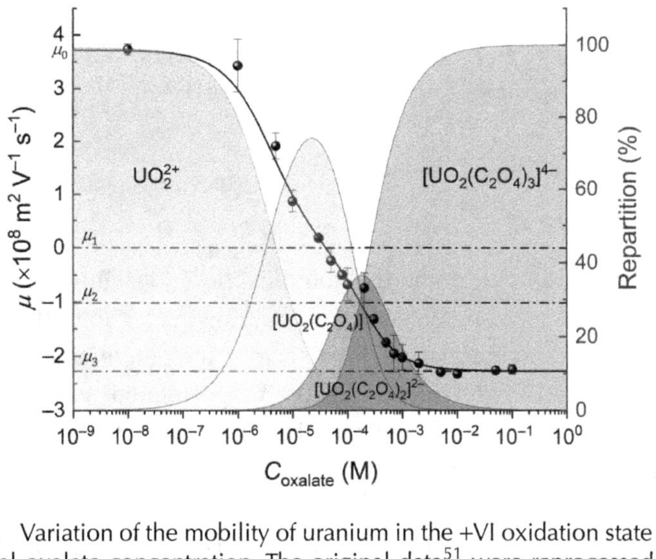

Figure 16. Variation of the mobility of uranium in the +VI oxidation state as a function of total oxalate concentration. The original data[51] were reprocessed using the rules given in Tables 4 and 5. Separation conditions: $V = 20$ kV, $L = 70$ cm, $\varnothing_{internal}$ = 50 μm, supporting electrolyte: NH_4Cl, $I = 0.1$ M, pH = 3.5. During the mathematical adjustment the electrophoretic mobilities were set to be constant for $\mu_0 = 3.72 \times 10^{-8}$ m^2 V^{-1} s^{-1} (UO_2^{2+}), $\mu_1 = 0$ m^2 V^{-1} s^{-1} ($[UO_2(C_2O_4)]$ is neutral), and $\mu_3 = -2.27 \times 10^{-8}$ m^2 V^{-1} s^{-1} ($[UO_2(C_2O_4)_3]^{4-}$), while the value of $\mu_2 = -1.27 \times 10^{-8}$ m^2 V^{-1} s^{-1} was estimated using the relationship $(\mu_2 - \mu_3)/\Delta z = 0.5 \times 10^{-8}$ m^2 V^{-1} s^{-1}. Under these conditions, the conditional stability constants obtained were (Eq. 33): log β^*_1 = 5.39 ± 0.10, log β^*_2 = 9.21 ± 0.09, and log β^*_3 = 12.89 ± 0.04. The resulting repartition diagram for the species is superimposed on the electrophoretic mobility variations, thereby reflecting, in these operating conditions, the proportions of species present for a given ligand concentration.

$$UO_2^{2+} + nC_2O_4^{2-} \rightleftharpoons [UO_2(C_2O_4)_n]^{(2n-2)-} \quad \beta_n^* = \frac{[[UO_2(C_2O_4)_n]^{(2n-2)-}]}{[UO_2^{2+}][C_2O_4^{2-}]^n} \tag{33}$$

The original data were reprocessed using these rules, which were not yet well established at the time of publication of the article.[51] The reduced number of variables to be determined naturally led to a decrease in the uncertainties associated with the stability constants. To do this, four of the seven variables were set to be constant during the mathematical adjustment process. Figure 16 shows two plateaus whose values correspond to the pure species UO_2^{2+} ($\mu_0 = 3.72 \times 10^{-8}$ m^2 V^{-1} s^{-1}) and $[UO_2(C_2O_4)_3]^{4-}$ ($\mu_3 = -2.27 \times 10^{-8}$ m^2 V^{-1} s^{-1}). Since the $[UO_2(C_2O_4)]$ complex is neutral, $\mu_1 = 0$ m^2 V^{-1} s^{-1}. For the $[UO_2(C_2O_4)_2]^-$ species, the relationship in Table 4 was applied, *i.e.*, $(\mu_2 - \mu_3)/2 = 0.5 \times 10^{-8}$ m^2 V^{-1} s^{-1}, providing $\mu_2 = -1.27 \times 10^{-8}$ m^2 V^{-1} s^{-1}.

2. *Isotachophoresis*

The method used to determine the complexation constants is identical to that used in capillary zone electrophoresis. However, due to the detection system employed (potential gradient detector), the potential gradient ratio, which is measured experimentally, must be correlated with the electrophoretic mobilities of the complexes being studied. Several qualitative indexes have been proposed in the literature;[53] we have opted to use the one devised by Hirokawa, namely R_E.[14] Thus, at isotachophoretic equilibrium (the stationary state), the ratio of the potential gradient of a zone containing an analyte A (E_A) to that of the leader electrolyte (E_L) can be related to the effective electrophoretic mobilities ($\overline{\mu_A}$ and $\overline{\mu_L}$) using the following relationship (Eq. 34),

$$R_E = \frac{E_A}{E_L} = \frac{\overline{\mu_A}}{\overline{\mu_L}} = \frac{h_A}{h_L} \tag{34}$$

where h_A and h_L are the step heights that can be measured directly. The effective mobilities are calculated using Eq. 28. For comparison and from

Table 6. List of studies dealing with the determination of stability constants (labile complexes) using isotachophoresis.

Cation	Ligand	Leader electrolyte	Terminal electrolyte	Ref.
Ba^{2+}, Ca^{2+}, Zn^{2+}, Cu^{2+}	Acetate	KOH (pH 5.0 by acetic acid)	5 mM TRIS (pH ≈ 5)	54
Ca^{2+}	Tartrate	10 mM HCl (pH 6.0 by histidine)	5 mM MES	
Mg^{2+}, Ca^{2+}, Sr^{2+}, Ba^{2+}, Cd^{2+}, Co^{2+}	Tartrate	KOH (pH 5.1 by tartric acid)	5 mM ε-AMC (pH 4.5)	55
Ba^{2+}, Sr^{2+}, Ca^{2+}, Mg^{2+}, Cd^{2+}, Co^{2+}, Ni^{2+}, Zn^{2+}	Glycolate, lactate, β-HB, α-HB, α-HIB	KOH (pH 4.4 or 5.0 by the ligand)	5 mM ε-AMC (pH 4.5)	56
Lanthanides	Acetate α-HIB	5–25 mM KOH (pH 4.8 by the ligand)	5 mM ε-AMC (pH 4.5)	57
Lanthanides, Y^{3+}	2-Ethyl-*n*-butyrate, malonate	20 mM NH_3 (pH 4.8 by the ligand)	20 mM carnitine hydrochloride	58
K^+, NH_4^+, Na^+	18-crown-6	10 mM HCl	10 mM TRIS (pH 8.5 by HCl) 10 mM TBACl	59

TRIS = tris(hydroxymethyl)aminomethane; MES = 2-(*N*-morpholino)ethanesulfonic acid; ε-AMC = ε-aminocaproic acid; β-HB = β-hydroxybutyrate; α-HB = α-hydroxybutyrate; α-HIB = α-hydroxyisobutyrate; TBACl = tetrabutylammonium chloride.

an experimental viewpoint, in capillary zone electrophoresis, the variation in migration time is the key parameter for evaluating labile complexes; in isotachophoresis, the exact equivalent is the variation in step height.

Very few studies in the literature deal with determining the stability constant using isotachophoresis. A compilation of these studies is provided in Table 6.

IX. Topological Speciation

This process involves a speciation in which the complexation reaction between a metallic chemical element and a ligand with a number of active sites is accompanied by a change in shape of the ligand depending on the site that is interacting with the metallic cation. This type of speciation mainly concerns proteins (such as transferrin, for example) that have affinities for metal cations. Transferrin is a glycoprotein with a molecular weight of nearly 80 000 Daltons that has two binding sites for iron, which are referred to as the C-lobe and the N-lobe. These sites are also sensitive to other metals that happen to be toxic to the organism, such as lead, cadmium, and plutonium. The complexation of the two sites by a metal leads to the closure of the two lobes, and consequently to a conformational change and a variation of the isoelectric point (*pI*) of the protein. In order to identify these conformational changes, the detection of the variation of the *pIs* can be performed using a special speciation method: capillary isoelectric focusing (cIEF). This technique, when coupled to an inductively coupled plasma mass spectrometer (cIEF-ICPMS), specifically enables the study of extremely toxic metals such plutonium or thorium at very low concentrations (10^{-9} M), which previously had been impossible. Historically, the first studies focused on the Fe^{III}/transferrin system.[16] Two examples of isoform separation of the Fe-transferrin complexes by means of UV detection (cIEF-UV) and of In-transferrin by means of ICPMS detection are respectively given in Figure 17 and Figure 18. Indium replaces the iron in cIEF-ICPMS, because the latter element is more strongly interfered by Ar in the plasma due to the presence of ArO^+, which has the same mass as the main iron isotope ($^{56}Fe^+$). It is interesting to note that despite the different size of the cations Fe^{3+} (r_i = 0.645 Å, CN = 6) and In^{3+} (r_i = 0.800 Å, CN = 6), complexation with transferrin leads to identical isoelectric point values within the limits of uncertainty. Because the internalization of iron into the cells requires the membrane receptors (transferrin receptors) to recognize only the holo T*f* form, *i.e.*, the 2:1 complex (Fe_cT*f* Fe_N), it seems important to determine if the closure of the lobes is effective for heavy metals and particularly for highly toxic elements such as plutonium. A recent study conducted on thorium and plutonium has highlighted the atypical behavior of the actinides.[60] A characteristic example of the behavior of the tetravalent actinides (Th and Pu) is shown in Figure 19. In this experiment, a Th^{4+} solution stabilized with nitrilotriacetic acid was reacted with an indium-loaded transferrin solution in order to partially replicate the conditions encountered in the blood, in which several iron complexes coexist. It can

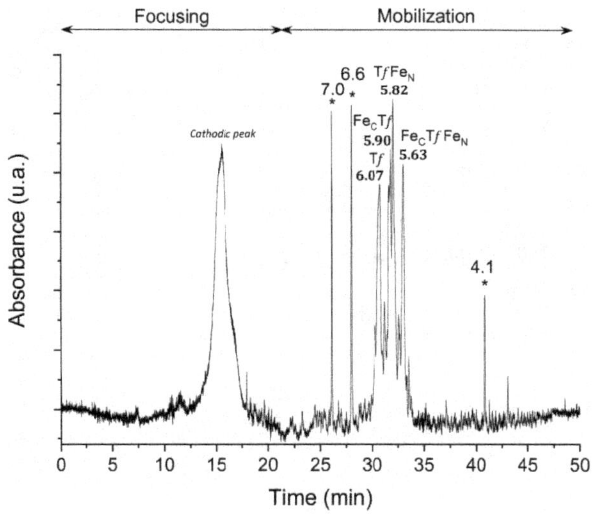

Figure 17. Separation of the iron-human transferrin complexes by cIEF and inverse UV detection. Conditions: 25 kV focusing, 20 min; 30 kV mobilization, 15 min; UV detection at 280 nm. (*) Internal standards used to calibrate the *pI* response. Two 1:1 iron complexes and one 1:2 iron complex coexist depending on whether the iron atom is located on the C-lobe or on the N-lobe, or on both lobes at once. Each complex has its own distinct isoelectric point, the *pI* values are indicated below the labels.

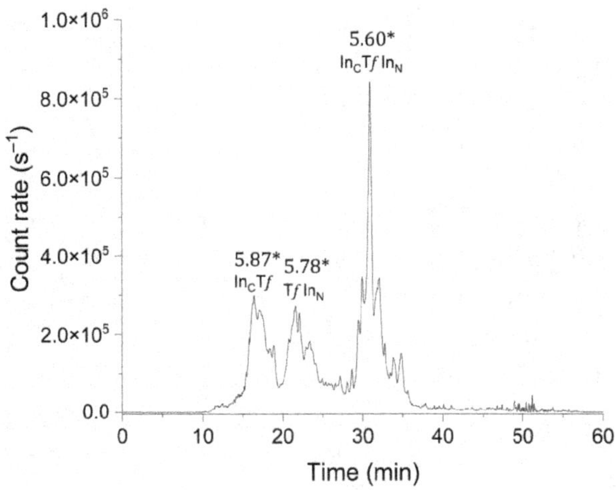

Figure 18. Separation of the indium-human transferrin complexes by cIEF and ICPMS detection at mass 115 (^{115}In). Conditions: $C_{Tf} = 50\ \mu M$, $C_{In} = 100\ \mu M$; 20 kV focusing, 45 min; 20 kV mobilization, 60 min at 1 psi. [arginine] = 1 mM used as a spacer (to prevent the separation electrolyte from leaving the capillary on the cathode side). (*) *pI* determined by a separate cIEF-UV experiment, under the same operating conditions. Adapted from Ref. 60.

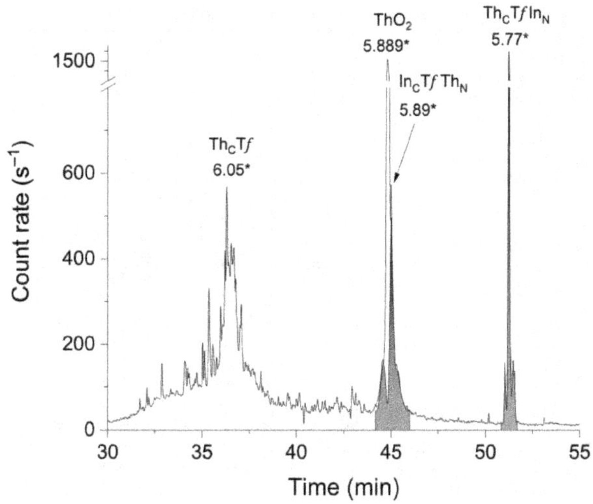

Figure 19. Separation of the isoforms of the thorium-human transferrin complexes by cIEF-ICPMS at mass 232 (^{232}Th). The thorium was placed in contact with a transferrin solution containing apotransferrin (*pI* = 6.05) and T*f*-In complexes (In^{3+} was taken as an analogue of Fe^{3+}) in which the indium is located either on the C-lobe (*pI* = 5.89) or on the N-lobe (*pI* = 5.77). Due to the presence of an alkaline region in the capillary because of the pH gradient, a part of thorium is precipitated as oxo-hydroxydo particles (summarized as ThO_2 by convenience). Note that the complexation with thorium does not modify the isoelectric point of the protein, which suggests that the lobes do not close in the presence of ions larger than indium despite a proven complexation. (*) *pI* determined by a separate cIEF-UV experiment, under the same operating conditions. Adapted from Ref. 60.

be seen that the thorium is complexed not only by the free form of the protein (apotransferrin), but also by the metalated forms. In each complex detected (Th_CT*f*, In_CT*f* Th_N and Th_CT*f* In_N), the isoelectric point found respectively corresponds to the species ApoT*f*, In_CT*f* and T*f* In_N. This type of result strongly suggests that even though the complexation is effective, it does not induce the closure of the lobes. Identical behavior was found for plutonium. The crystallographic radii of these elements are larger than those of Fe^{3+} and In^{3+} ($r_{Th^{4+}}$ = 0.94 Å, $r_{Pu^{4+}}$ = 0.86 Å, CN = 6). It is interesting to note that the size of the indium ion is not very different from that of plutonium; however, the former induces a closure of the lobe, while the latter does not.

X. Special Cases

As mentioned previously, electromigration techniques alone cannot unequivocally determine the nature of the complexes, because the only observable feature that can be measured is a speed. The key is therefore to relate these speed variations to firm hypotheses. In the examples provided above, it is easy to obtain information about the speciation of the metals, because the operating conditions chosen have made it possible to restrict the number of possible solutions. There are unfortunately some cases in which determining the speciation is very difficult. In such cases, complementary techniques (or independent information) are required to elucidate the system. A few typical examples are given below.

A. Protonated Complexes

Protonated metal-ligand complexes are the most challenging systems to study with capillary electrophoresis. Indeed, the domain of existence of these complexes depends on the pH selected to carry out the separation. Without complementary information, it is impossible to postulate that a certain complex exists on the basis of electrophoretic mobility measurements alone. An accurate assessment would require systematic research such as in the example shown in Figure 20 for several pH values. While such a systematic approach is not impossible to implement, it remains very time-consuming. For example, to study the complexation of scandium (Sc^{3+}) by DTPA, a series of four studies was performed by CE–ICP-MS at pH values 1.35, 1.85, 2.11 and 2.94.[61] Mathematical adjustment of the variations of the observed mobility of scandium containing species (see Figure 20, left) enabled the determination of four conditional constants. Complementary experiments were carried out at pH 1.05 by ^{45}Sc nuclear magnetic resonance (NMR) and at pH 1.40 by free-ion selective radiotracer extraction. The variation of the conditional constant as a function of proton concentration ($p_CH = -\log [H^+]$) is shown on the right-hand side of Figure 20. Two slopes are determined, with their difference being attributed to a modification of the charge of the complex, likely according to the following equilibrium (Eq. 35):

$$\left[Sc(DTPA)H\right]^{-} \rightleftharpoons \left[Sc(DTPA)\right]^{2-} + H^{+}, \ \log K \approx 2 \qquad (35)$$

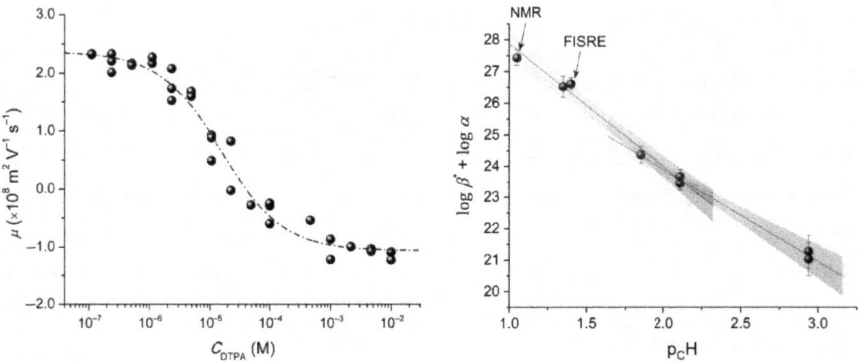

Figure 20. (left) Electrophoretic mobility of scandium as a function of total DTPA concentration. Separation conditions: T = 298 K, supporting electrolyte: NaCl, I = 0.1 M, pH = 1.35. Mathematically adjusting the points using Eq. 28 allows a constant log β^* = 26.52 ± 0.34 to be determined. (right) Variation of the conditional stability constants corrected for the dissociation of the acid (log α) as a function of proton concentration ($p_CH = -\log [H^+]$) determined by CE–ICP-MS, ^{45}Sc NMR, and free-ion selective radiotracer extraction (FISRE).[61] Adapted from Ref. 61.

The previous case involved a polyacid whose limiting species is a 1:1 complex. Yet there exist much more complicated systems involving polyacids whose maximal order can be as high as three. For instance, this is the case of the Am^{3+}/Cit^{3-} (Cit^{3-} = citrate) system, for which third-order complexes are observed. In acidic medium, species such as [Am(HCit)], [Am(H_2Cit)(H_2Cit)], [Am(HCit)(H_2Cit)], *etc.*, can therefore coexist. In such a situation, electromigration techniques are useless unless they are used in conjunction with other techniques. For these truly complicated and multifactorial cases, the relevant piece of information provided by electromigration techniques is the sign of the charge carried by the complex, particularly when the sign of the mobility changes, which is evidence for the presence of a higher complex. This strategy was successfully employed by Tamain *et al.*[62] CE–ICP-MS coupling was used in conjunction with EXAFS experiments, DFT calculations and spectroscopic techniques (time-resolved laser-induced fluorescence, UV-vis and ^{13}C NMR). In this study, CE–ICP-MS was used only to determine the charge of the complexes at the two pH values considered (pH 1 and 3). It was thus

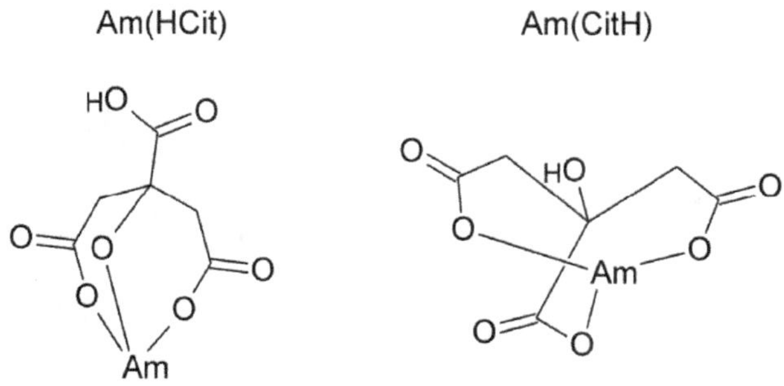

Figure 21. Proposed structures for the different Am^{3+} 1:1 complexes formed with citrate prevailing at pH 1 ([Am(HCit)], left) and 3 ([Am(CitH)], right).

established that at pH 3, the 1:1 and 1:2 complexes are neutral, while the 1:3 complex is neutral or negative. At pH 1, the first two complexes are neutral. Processing the correlations found among all of these techniques made it possible to propose, for example, the following structures at pH 3: [Am(CitH)] for the 1:1 complex and [Am(HCitH)(H_2CitH)] for the 1:2 complex. At pH 1 for example, the 1:1 complex is [Am(HCit)] and not [Am(CitH)]; in the former complex, the proton is located on the central carboxylic group, but in the latter it is located on the hydroxyl group in the alpha position (Figure 21).[62]

B. Invisible Complexes

This type of situation is exemplified by the Th^{4+}/NTA^{3-} (nitrilotriacetate) system.[18] Prior to the publication of this study, only the 1:1 complex had been proved to exist, and its complexation constant was well known: log $\beta_1 = 16.9$.[63] The 1:2 complex was thought to exist but proved to be difficult to detect through the usual techniques such as spectrophotometry or potentiometry. CE–ICP-MS coupling showed that electrophoretic mobility became negative for NTA concentrations above 10^{-3} M, thereby unambiguously demonstrating the existence of the $[Th(NTA)_2]^{2-}$ complex. The determination of this constant requires that the value of the constant β_1 is taken into account, as will be shown below. For this system, the following chemical equilibria (Eqs. 36 and 37) must be considered.

$$\mathrm{Th^{4+} + NTA^{3-} \rightleftharpoons [Th(NTA)]^{+}} \qquad \beta_1 = \frac{\left[\mathrm{[Th(NTA)]^{+}}\right]}{\left[\mathrm{Th^{4+}}\right]\left[\mathrm{NTA^{3-}}\right]} \tag{36}$$

$$\mathrm{Th^{4+} + 2NTA^{3-} \rightleftharpoons [Th(NTA)_2]^{2-}} \qquad \beta_2 = \frac{\left[\mathrm{[Th(NTA)_2]^{2-}}\right]}{\left[\mathrm{Th^{4+}}\right]\left[\mathrm{NTA^{3-}}\right]^2} \tag{37}$$

The following conventions are set out for the sake of clarity: $\mu_0 = \mu_{\mathrm{Th^{4+}}}$, $\mu_1 = \mu_{\mathrm{[Th(NTA)]^{+}}}$ and $\mu_2 = \mu_{\mathrm{[Th(NTA)_2]^{2-}}}$.

The variation (and sign) of the electrophoretic mobility as a function of ligand concentration generally enables firm hypotheses to be put forth regarding the number and type of complexes. Thus, for NTA concentrations between 10^{-10} and 10^{-6} M, Figure 22 shows a plateau that makes it possible to postulate the existence of a predominant species, which in this case is naturally $\mathrm{Th^{4+}}$. Between 10^{-3} and 10^{-2} M, the electrophoretic

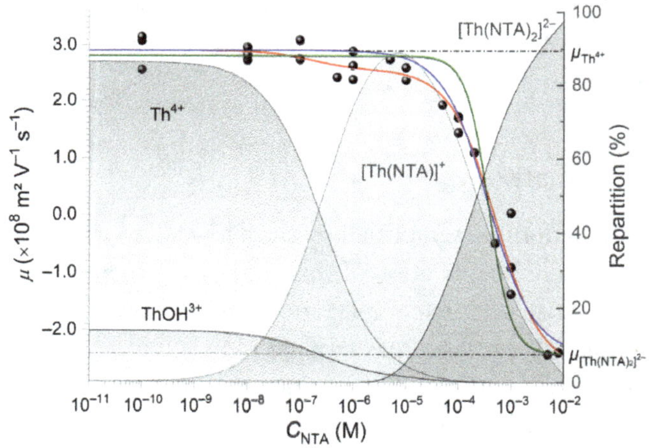

Figure 22. Left axis: Variation of the mobility of thorium as a function of nitrilotriacetic acid concentration at p_CH 1.395. Blue curve: mathematical adjustment using a sigmoid function (Eq. 38), green curve: mathematical adjustment (Eq. 40) taking into account the existence of only the predominant 1:2 complex, red curve: mathematical adjustment taking into account the existence of the two 1:1 and 1:2 complexes (Eq. 41). Right axis: Repartition of the species under the conditions of the experiment: $C_{\mathrm{Th}} = 10^{-7}$ M, $p_CH = 1.395$, $10^{-11} < C_{\mathrm{NTA}} < 10^{-2}$ M.

mobility is negative, which allows us to suppose that the main species is $[Th(NTA)_2]^{2-}$. This hypothesis is reinforced by the fact that the last two points on the right in Figure 22 correspond to identical mobility values. This means that the plateau has likely been reached and that only the limiting complex is present. It is also worth noting that the variation of electrophoretic mobility from a positive value to a negative value is extremely abrupt. Consequently, the formation constant is probably very large, because the transition from Th^{4+} to $[Th(NTA)_2]^{2-}$ is sharp and the intermediate species $[Th(NTA)]^{+}$ ought to remain of minor importance. The principle of minimizing the number of parameters therefore leads us to consider only two species in the adjustment of the experimental data.

When processing the experimental data shown in Figure 22, two methods of minimization can be applied, with both of them yielding a complexation constant whose value is consistent within experimental uncertainty. The first method consists in adjusting the experimental data points using a logistic sigmoid function. As a matter of fact, it can be easily demonstrated that the variation of the electrophoretic mobilities follows this type of rule (Eq. 38),

$$\mu = \mu_b + \frac{\mu_a - \mu_b}{1 + \left(\frac{C_{NTA}}{C_{NTA}^0}\right)^p} \tag{38}$$

with $\mu_a = \mu_0$, $\mu_b = \mu_2$, and $p = 1$; this last component can be controlled or set to a fixed value. This method is preferable, because there is only one unknown parameter (C_{NTA}^0). In the example given, an initial iteration led to $p = 0.90$. This value is close to 1, which means that the values obey the expected rule and that apparently only two species are present. For the second iteration, therefore, this value was set to 1 (see blue curve in Figure 22). At the inflexion point of this function, the inverse of the concentration found gives direct access to the value of the complexation constant under these experimental conditions. To obtain the apparent complexation constant, it is necessary to add a term that makes it possible to move from the total NTA concentration (C_{NTA}) to the free ligand equilibrium concentration ($[NTA^{3-}]$) (Eq. 39).

$$\alpha = \frac{C_{\mathrm{NTA}}}{[\mathrm{NTA}^{3-}]} = 1 + 10^{pKa_4}[\mathrm{H}^+] + 10^{(pKa_4+pKa_3)}[\mathrm{H}^+]^2 + 10^{(pKa_4+pKa_3+pKa_2)}[\mathrm{H}^+]^3 + 10^{(pKa_4+pKa_3+pKa_2+pKa_1)}[\mathrm{H}^+]^4 \quad (39)$$

Under the acidic conditions of the experiment (p_CH = 1.395), this term expressed in log unit is: log $\alpha = 10.08 \pm 0.09$ for β_1 and $2 \times \log \alpha$ for β_2.[18] The result is log $\beta_2 = 23.61 \pm 0.33$. This finding may seem correct at first glance. Nevertheless, it means that the constant for the equilibrium $[\mathrm{Th(NTA)}]^+ + \mathrm{NTA}^{3-} \rightleftharpoons [\mathrm{Th(NTA)_2}]^{2-}$ is log K_2 = 23.61 – 16.9 = 6.71. This result is surprising, because one generally expects to find a value of the same magnitude if no steric hindrance opposes the formation of the higher complex.

One can check this result by applying the second adjustment equation (Eq. 40), which, contrary to Eq. 38, explicitly expresses the complexation constant. As in Eq. 38, only one unknown needs to be determined:

$$\mu = \frac{\mu_0 + \mu_2 \times \beta_2 \times \left({C_{\mathrm{NTA}}}/{\alpha} \right)^2}{1 + \beta_2 \times \left({C_{\mathrm{NTA}}}/{\alpha} \right)^2} \quad (40)$$

Attention must be paid to this equation which is valid only if the total metal concentration is negligible compared to that of the ligand. Otherwise, the free ligand equilibrium concentration needs to be recalculated for these few points by solving iteratively the mass balance equations for both the metal and ligand at each refinement step of the equilibrium constants (*e.g.* β_2 in the case of Eq. 40). The green curve in Figure 22 is the outcome of the mathematical adjustment using Eq. 40 and log $\beta_2 = 27.03 \pm 0.34$. It can be noticed that Eq. 40 does not properly fit the experimental data points, especially around the inflection point, which is the critical area for determining the complexation constant. Furthermore, this result differs substantially from the previous one, even though the two curves (blue and green) meet at the same inflection point.

These divergent results between the two methods suggest the existence of an additional species. Since the intermediate species $[Th(NTA)]^+$ is known to exist, Eq. 40 must be modified. The equation becomes (Eq. 41):

$$\mu = \frac{\mu_0 + \mu_1 \times \beta_1 \times {C_{NTA}}/{\alpha} + \mu_2 \times \beta_2 \times \left({C_{NTA}}/{\alpha}\right)^2}{1 + \beta_1 \times {C_{NTA}}/{\alpha} + \beta_2 \times \left({C_{NTA}}/{\alpha}\right)^2} \quad (41)$$

The divergent values obtained by considering only one complex indicate that the 1:1 complex is, in a sense, invisible. Because the 1:1 constant is well known, it is given a set value during the mathematical adjustment of the points using Eq. 41. The outcome is the red curve in Figure 22. The value of the constant is finally: $\log \beta_1 = 30.73 \pm 0.22$.[18] It is important to note that the red curve is one solution among others. Here, $\mu_1 = 2.6 \times 10^{-8}$ $m^2\,V^{-1}\,s^{-1}$. By constraining this parameter such that, for example, $\mu_1 = \mu_0$, we reach an identical result for β_2. Under these operating conditions (p_CH = 1.395), the predominant form of the acid is the neutral form $(NTA)H_3$ (70%). One must absolutely not attribute this mobility, whose possible values are $\approx 2.5 \times 10^{-8} < \mu_1 < \mu_0 = 2.9 \times 10^{-8}$ $m^2\,V^{-1}\,s^{-1}$ to the existence of a complex such as $[Th(NTA)H_3]^{4+}$. In fact, the 1:1 complex is "invisible" from an electrophoretic mobility standpoint even though it predominates, as can be seen in the speciation diagram in Figure 22; its presence influences only the denominator of Eq. 38, as it reduces the slope of the curve. This observation is open to more than one, not necessarily exclusive, interpretation: a) the $[Th(NTA)]^+$ complex exhibits chaotropic behavior (its mobility is higher than expected for a monocharged complex, because it destructures the water in its vicinity); and b) the electrophoretic mobility of highly charged ions such as Th^{4+} decreases more rapidly with increasing ionic strength than that of monocharged ions such as $[Th(NTA)]^+$, and can become identical where ionic strength conditions allow. For the latter hypothesis, an approximate retroactive calculation can be performed using the Fuoss-Onsager equation (Eq. 3) in order to determine the mobilities at zero ionic strength. It should be recalled that Eq. 3

is less accurate for representing the behavior of highly charged ions (+3 and +4) as compared to monovalent or divalent ions. Under the experimental conditions employed, the electrophoretic mobility of thorium at zero ionic strength (omitting $[Th(OH)]^{3+}$) is found to be equal to $\mu_0 \approx 8 \times 10^{-8}$ m^2 V^{-1} s^{-1}, which deviates from the experimental value by only 8%: $\mu_0 = (7.1 \pm 0.3) \times 10^{-8}$ m^2 V^{-1} s^{-1}.[64] For the $[Th(NTA)]^+$ complex, the following value is obtained: $\mu_0 \approx 4 \times 10^{-8}$ m^2 V^{-1} s^{-1}. The second hypothesis therefore seems to be the more appropriate one. Therefore, it is simply due to chance that identical electrophoretic mobilities were determined for ions with very different charges.

XI. General Conclusions

Electromigration techniques are non-intrusive techniques that preserve chemical species. Thus, they make it possible to study them and determine their chemical stability. However, they are tricky to use because of the large number of parameters (temperature, ionic strength, *etc.*) that need to be strictly controlled. It is not surprising, therefore, to find, after all, only few publications describing the determination of stability constants by these methods. However, when coupled with highly sensitive detectors such as mass spectrometers, they become valuable tools for studying chemical systems that are extremely difficult to investigate by other techniques. Thus, for example, plutonium at the oxidation state +V (PuO_2^+) is unstable and rapidly disproportionates into other oxidation states. Since this reaction is dependent on the concentration of the metal (the kinetics of this reaction decreases by a factor of 10^n if the concentration decreases by the same amount), the use of a mass spectrometer makes it possible to study this system at concentrations at which disproportionation reactions are virtually stopped. The coupling between a capillary electrophoresis instrument and an ICP-MS is to our knowledge the only available tool capable of unravelling the speciation of plutonium at the oxidation state +V in order to rationalize its fate and behavior in the environment.

These coupled techniques have also significant potential in the field of biochemistry. For example, half of all proteins expressed in the human body contain a metal, which can be detected by a mass spectrometer. The example given in this chapter for the transferrin/actinide system illustrates this potential. It gave rise to an additional definition of speciation called by the authors "topological speciation". In addition, unexpected results obtained by CE–ICP-MS have shown that electromigration techniques have the potential to probe the structure of complexes and to determine bond length with an accuracy of about 0.05 Å.

However, there is still theoretical work to be done to quantify the electrophoretic mobilities of complexes and to uncover semi-empirical or even better theoretical relationships between them. The challenge is to determine with great precision the stability of complexes between metals and ligands, but to date, it is often difficult to estimate accurately the mobility values for the charged species. Despite the very limited number of data sets available, empirical correlations within homologous series could be established, showing proportionality relationships between the mobility and the electrical charge of ML_n-type complexes. However, more work is necessary to consolidate these findings and firmly establish practical rules.

XII. References

1. Onsager, L.; Fuoss, R. M. Irreversible processes in electrolytes. Diffusion, conductances, and viscous flow in arbitrary mixtures of strong electrolytes. *J. Phys. Chem.* **1932**, 36, 2689–2778.
2. Jaros, M.; Vcelakova, K.; Zuskova, I.; Gas, B. Optimization of background electrolytes for capillary electrophoresis: II. Computer simulation and comparison with experiments. *Electrophoresis* **2002**, 23, 2667–2677.
3. Gas, B. https://web.natur.cuni.cz/gas/peakmaster.html
4. Stedry, M.; Jaros, M.; Gas, B. Eigenmobilities in background electrolytes for capillary zone electrophoresis — I. System eigenpeaks and resonance in systems with strong electrolytes. *J. Chromatogr. A* **2002**, 960, 187–198.
5. Stedry, M.; Jaros, M.; Hruska, V.; Gas, B. Eigenmobilities in background electrolytes for capillary zone electrophoresis: III. Linear theory of electromigration. *Electrophoresis* **2004**, 25, 3071–3079.
6. Jaros, M.; Hruska, V.; Stedry, M.; Zuskova, I.; Gas, B. Eigenmobilities in background electrolytes for capillary zone electrophoresis: IV. Computer program PeakMaster. *Electrophoresis* **2004**, 25, 3080–3085.
7. Masselter, S. M.; Zemann, A. J. Influence of organic solvents in coelectroosmotic capillary electrophoresis of phenols. *Anal. Chem.* **1995**, 67, 1047–1053.
8. Melanson, J. E.; Baryla, N. E.; Lucy, C. A. Double-chained surfactants for semipermanent wall coatings in capillary electrophoresis. *Anal. Chem.* **2000**, 72, 4110–4114.

9. Finkler, C.; Charrel, H.; Engelhardt, H. Permanent coated capillaries with reversed electroosmotic flow for anion analysis. *J. Chromatogr. A* **1998**, 822, 101–106.
10. Foret, F.; Sustacek, V.; Bocek, P. On-line isotachophoretic sample preconcentration for enhancement of zone detectability in capillary zone electrophoresis. *J. Microcol. Sep.* **1990**, 2, 229–233.
11. Kohlrausch, F. Ueber Concentrations — Verschiebungen durch Electrolyse im Inneren von Lösungen und Lösungsgemischen. *Ann. Phys. Chem.* **1897**, 62, 209–239.
12. Beckers, J. L.; Everaerts, F. M.; Ackermans, M. T. Isotachophoresis with electroosmotic flow: open *versus* closed systems. *J. Chromatogr.* **1991**, 537, 429–442.
13. Melzer, T.; Wimmer, B.; Bock, S.; Posch, T. N.; Huhn, C. Challenges and applications of isotachophoresis coupled to mass spectrometry: a review. *Electrophoresis* **2020**, 41, 1045–1059.
14. Hirokawa, T. Chapter 8 — Capillary Isotachophoresis. In *Capillary Electromigration Separation Methods*, Poole, C. F., Ed. Elsevier: Amsterdam, 2018.
15. Shimura, K. Chapter 7 — Capillary Isolectric Focusing. In *Capillary Electromigration Separation Methods*, Poole, C. F., Ed. Elsevier: Amsterdam, 2018.
16. Kilar, F.; Hjerten, S. Fast and high resolution analysis of human serum transferrin by high performance isoelectric focusing in capillaries. *Electrophoresis* **1989**, 10, 23–29.
17. Aupiais, J.; Chartier, F. Capillary electrophoresis. In *Sample Introduction Systems in ICPMS and ICPOES*, Beauchemin, D., Ed. Elsevier: Amsterdam, 2020.
18. Bonin, L.; Aupiais, J.; Kerbaa, M.; Moisy, P.; Topin, S.; Siberchicot, B. Revisiting the actinide DTPA complexes in aqueous solution by CE-ICPMS and ab initio molecular dynamics. *RSC Adv.* **2016**, 6, 62729–62741.
19. Aupiais, J.; Bonin, L.; den Auwer, C.; Moisy, P.; Siberchicot, B.; Topin, S. On the use of speciation techniques and *ab initio* modelling to understand tetravalent actinide behavior in biological medium: An^{IV}DTPA case. *Dalton Trans.* **2016**, 45, 3759–3770.
20. Aupiais, J. Optimization of cation detection by capillary zone electrophoresis with indirect UV absorbance. *Chromatographia* **1997**, 44, 303–312.
21. Weston, A.; Brown, P. R. Factors affecting the separation of inorganic metal cations by capillary electrophoresis. *J. Chromatogr.* **1992**, 593, 289–295.
22. Weston, A.; Brown, P. R.; Heckenberg, A. L.; Jandik, P.; Jones, W. R. Effect of electrolyte composition on the separation of inorganic metal cations by capillary ion electrophoresis. *J. Chromatogr.* **1992**, 602, 249–256.
23. Olesik, J. W. Capillary electrophoresis for elemental speciation studies. In *Comprehensive Analytical Chemistry*, Caruso, J. A.; Sutton, K. L.; Ackley, K. L., Eds. Elsevier: 2000.
24. Timerbaev, A. R. Element speciation analysis by capillary electrophoresis. *Talanta* **2000**, 52, 573–606.
25. Hirokawa, T.; Nishimoto, K.; Nishiyama, F. Isotachophoretic separation of Fe(II) and Fe(III) by using 1,10-phenantroline as a complex-forming agent. *J. Chromatogr. A* **1996**, 723, 389–394.

26. Liu, Y.; Lopez-Avila, V.; Zhu, J. J.; Wiederin, D. R.; Beckert, W. F. Capillary electrophoresis coupled on-line with inductively coupled plasma mass spectrometry for elemental speciation. *Anal. Chem.* **1995**, 67, 2020–2025.
27. Dabek-Zlotorzynska, E.; Lai, E. P. C.; Timerbaev, A. R. Capillary electrophoresis: the state-of-the-art in metal speciation studies. *Anal. Chim. Acta* **1998**, 359, 1–26.
28. Timerbaev, A. R.; Timerbaev, R. M. Recent progress of capillary electrophoresis in studying the speciation of actinides. *Trends Anal. Chem.* **2013**, 51, 44–50.
29. Timerbaev, A. R. Element Speciation Analysis Using Capillary Electrophoresis: Twenty Years of Development and Applications. *Chem. Rev.* **2013**, 113, 778–812.
30. Ali, I.; Gupta, V. K.; Aboul-Enein, H. Y. Metal ion speciation and capillary electrophoresis: Application in the new millennium. *Electrophoresis* **2005**, 26, 3988–4002.
31. Guo, N.; Pottier, F.; Aupiais, J.; Alliot, C.; Montavon, G.; Champion, J. Evidence for the heaviest expected halide species in aqueous solution, At^-, by electromobility measurements. *Inorg. Chem.* **2018**, 57, 4926–4933.
32. Zhou, D.; Lin, Y.; Long, H.; Xu, Y.; Wang, B.; Xian, L.; Xia, C.; Hou, X.; Zheng, C. Simultaneous total and speciation analysis of rhenium by capillary electrophoresis-inductively coupled plasma mass spectrometry. *Spectrochim. Acta B* **2021**, 180, 106211.
33. Martelat, B.; Isnard, H.; Vio, L.; Dupuis, E.; Cornet, T.; Nonell, A.; Chartier, F. Precise U and Pu isotope ratio measurements in nuclear samples by hyphenating capillary electrophoresis and MC-ICPMS. *Anal. Chem.* **2018**, 90, 8622–8628.
34. Graser, C. H.; Banik, N. L.; Bender, K. A.; Lagos, M.; Marquardt, C. M.; Marsac, R.; Montoya, V.; Geckeis, H. Sensitive Redox Speciation of Iron, Neptunium, and Plutonium by Capillary Electrophoresis Hyphenated to Inductively Coupled Plasma Sector Field Mass Spectrometry. *Anal. Chem.* **2015**, 87, 9786–9794.
35. Hanf, L.; Henschel, J.; Diehl, M.; Winter, M.; Nowak, S. Mn^{2+} or Mn^{3+}? Investigating transition metal dissolution of manganese species in lithium ion battery electrolytes by capillary electrophoresis. *Electrophoresis* **2020**, 41, 697–704.
36. David, F.; Vokhmin, V. Thermodynamic properties of some tri- and tetravalent actinide aquo ions. *New J. Chem.* **2003**, 27, 1627–1632.
37. Leguay, S.; Vercouter, T.; Topin, S.; Aupiais, J.; Guillaumont, D.; Miguirditchian, M.; Moisy, P.; Le Naour, C. New Insights into Formation of Trivalent Actinides Complexes with DTPA. *Inorg. Chem.* **2012**, 51, 12638–12649.
38. Brown, A.; Paulenova, A.; Gelis, A. V. Aqueous complexation of thorium(IV), uranium(IV), neptunium(IV), plutonium(III/IV), and cerium(III/IV) with DTPA. *Inorg. Chem.* **2012**, 51, 7741–7748.
39. Clark, D. L.; Conradson, S. D.; Webster Keogh, D.; Neu, M. P.; Palmer, P. D.; Runde, W.; Scott, B. L.; Tait, C. D. In *X-ray absorption and diffraction studies of monomeric actinide tetra-, penta-, and hexavalent carbonato complexes*, Workshop on Speciation, Techniques, and Facilities for Radioactive Materials at Synchrotron Light Sources, Grenoble, 1998; OECD Publications: Grenoble, 1998.

40. Conradson, S. D.; Abney, K. D.; Begg, B. D.; Brady, E. D.; Clark, D. L.; den Auwer, C.; Ding, M.; Dorhout, P. K.; Espinosa-Faller, F. J.; Gordon, P. L.; Haire, R. G.; Hess, N. J.; Hess, R. F.; Webster Keogh, D.; Lander, G. H.; Lupinetti, A. J.; Morales, L. A.; Neu, M. P.; Palmer, P. D.; Paviet-Hartmann, P.; Reilly, S. D.; Runde, W. H.; Tait, C. D.; Veirs, D. K.; Wastin, F. Higher order speciation effects on plutonium L_3 X-ray absorption near edge spectra. *Inorg. Chem.* **2004**, 43, 116–131.
41. Clark, D. L.; Conradson, S. D.; Ekberg, S. A.; Hess, N. J.; Neu, M. P.; Palmer, P. D.; Runde, W.; Tait, C. D. EXAFS studies of pentavalent neptunium carbonato complexes. Structural elucidation of the principal constituents of neptunium in groundwater environments. *J. Am. Chem. Soc.* **1996**, 118, 2089–2090.
42. Aupiais, J.; Alexandre, J. C.; Sicre, R.; Siberchicot, B.; Topin, S.; Moisy, P.; Dacheux, N. The Np^V and Pu^V carbonate systems: Thermodynamics and coordination chemistry. *Eur. J. Inorg. Chem.* **2020**, 216–225.
43. Topin, S.; Aupiais, J.; Moisy, P. Direct determination of plutonium(V) and neptunium(V) complexation by carbonate ligand with CE-ICP-sector field MS. *Electrophoresis* **2009**, 30, 1747–1755.
44. Sonke, J. E.; Salters, V. J. M. Disequilibrium effects in metal speciation by capillary electrophoresis inductively coupled plasma mass spectrometry (CE-ICP-MS); theory, simulations and experiments. *Analyst* **2004**, 129, 731–738.
45. Sauge-Merle, S.; Lemaire, D.; Evans, R. W.; Berthomieu, C.; Aupiais, J. Revisiting binding of plutonium to transferrin by CE-ICP-MS. *Dalton Trans.* **2017**, 46, 1389–1396.
46. Gebauer, P.; Bocek, P.; Deml, M.; Janak, J. Isotachophoresis of kinetically labile complexes. *J. Chromatogr.* **1980**, 199, 81–94.
47. Hirokawa, T.; Mao, Q. New method to determine stability constants of lanthanide-EDTA complexes by isotachophoresis. *J. Chromatogr. A* **1997**, 786, 377–381.
48. Koudelkova, M.; Vinsova, H.; Jedinakova-Krizova, V. Isotachophoretic determination of stability constants of Ho and Y complexes with diethylenetriaminepentaacetic acid and 1,4,7,10-tetraazadodecane-N,N',N",N"'-tetraacetic acid. *J. Chromatogr. A* **2003**, 990, 311–316.
49. Anderko, A.; Lencka, M. M. Computation of electrical conductivity of multicomponent aqueous systems in wide concentration and temperature ranges. *Ind. Eng. Chem. Res.* **1997**, 36, 1932–1943.
50. Chandler, E. E. The ionization constants of the second hydrogen ion of dibasic acids. *J. Am. Chem. Soc.* **1908**, 30, 694–713.
51. Petit, J.; Aupiais, J.; Topin, S.; Geertsen, V.; Beaucaire, C.; Stambouli, M. Stability constants determination of successive metal complexes by hyphenated capillary electrophoresis-inductively coupled plasma mass spectrometry. *Electrophoresis* **2010**, 31, 355–363.
52. Willberger, C.; Leichtfuß, D.; Amayri, S.; Reich, T. Determination of the stability constants of the acetate complexes of the actinides Am(III), Th(IV), Np(V), and

U(VI) using capillary electrophoresis-inductively coupled plasma mass spectrometry. *Inorg. Chem.* **2019**, 58, 4851–4858.

53. Hirokawa, T.; Nishino, M.; Aoki, N.; Kiso, Y.; Sawamoto, Y.; Yagi, T.; Akiyama, J.-I. Table of isotachophoretic indices — I. Simulated qualitative and quantitative indices of 287 anionic substances in the range pH 3–10. *J. Chromatogr.* **1983**, 271, D1-D106.
54. Kiso, Y.; Hirokawa, T. A new method for determination of complex stability constant by isotachophoresis. *Chem. Lett.* **1980**, 745–748.
55. Hirokawa, T.; Kiso, Y. Complex-forming equilibria in isotachophoresis. II. Evaluation of stability constants of tartrate and citrate complexes. *J. Chromatogr.* **1982**, 248, 341–362.
56. Hirokawa, T.; Matsuki, T.; Takemi, H.; Kiso, Y. Complex-forming equilibria in isotachophoresis. V. Estimation of coordination numbers and stability constants of metal-oxycarboxylic acid complexes by means of computer simulation. *J. Chromatogr.* **1983**, 280, 233–247.
57. Hirokawa, T.; Aoki, N.; Kiso, Y. Complex-forming equilibria in isotachophoresis. VI. Simulation of isotachophoretic equilibria of lanthanoids and determination of mobilities and stability constants of acetate and α-hydroxyisobutyrate complexes. *J. Chromatogr.* **1984**, 312, 11–29.
58. Hirokawa, T.; Hashimoto, Y. Simultaneous separation of yttrium and lanthanide ions by isotachophoresis. *J. Chromatogr. A* **1997**, 772, 357–367.
59. Stover, F. S. Isotachophoresis of metal-neutral ligand complexes: 18-crown-6 ether complexes with alkali metals. *J. Chromatogr.* **1984**, 298, 203–210.
60. Brulfert, F.; Aupiais, J. Topological speciation of actinide–transferrin complexes by capillary isoelectric focusing coupled with inductively coupled plasma mass spectrometry: evidence of the non-closure of the lobes. *Dalton Trans.* **2018**, 47, 9994–10001.
61. Huclier-Markai, S.; Alliot, C.; Sebti, J.; Brunel, B.; Aupiais, J. A comparative thermodynamic study of the formation of scandium complexes with DTPA and DOTA. *RSC Adv.* **2015**, 5, 99606–99617.
62. Tamain, C.; Bonato, L.; Aupiais, J.; Dumas, T.; Guillaumont, D.; Barkleit, A.; Berthon, C.; Solari, P. L.; Ikeda-Ohno, A.; Guilbaud, P.; Moisy, P. Role of the hydroxo group in the coordination of citric acid to trivalent americium. *Eur. J. Inorg. Chem.* **2020**, 2020, 1331–1344.
63. Anderegg, G. Critical survey of stability constants of NTA complexes. *Pure Appl. Chem.* **1982**, 54, 2693–2758.
64. Arndt, F.; Marx, G. Determination of ionic mobilities of plutonium(VI) in aqueous solutions at 298 K by use of conductivities. *Z. Naturforsch.* **1981**, 36a, 1019–1020.

https://doi.org/10.1142/9789811281099_0004

4 Thermo-Optical Measurement Methods to Study Biomolecular Interactions — Microscale Thermophoresis and More

Annemarie Lüdecke, Andreas Langer, Werner Streicher, Stefan Duhr and Philipp Baaske*

NanoTemper Technologies GmbH, Flößergasse 4,
81369 München, Germany
*Philipp.Baaske@nanotempertech.com

Table of Contents

I. List of Abbreviations

ATP adenosine triphosphate
CAS9 CRISPR associated protein 9
CRISPR clustered regularly interspaced short palindromic repeats
DNA deoxyribonucleic acid
DOL degree of labeling

DTT	dithiothreitol
EDTA	ethylenediaminetetraacetic acid
FCS	fluorescence correlation spectroscopy
FRET	Förster resonance energy transfer
GPCR	G-protein-coupled receptor
HEPES	4-(2-hydroxyethyl)-1-piperazineethanesulfonic acid
His-tag	polyhistidine-tag
HPLC	high-performance liquid chromotography
IR	infrared
ITC	isothermal titration calorimetry
LIFC	ligand induced fluorescence change
MST	microscale thermophoresis
NHS	*N*-hydroxysuccinimide
NTA	nitrilotriacetic acid
p38α	mitogen-activated protein kinase 14
PBP	phosphate binding protein
PBS	phosphate buffered saline
PROTAC	proteolysis targeting chimera
RNA	ribonucleic acid
SDS	sodium dodecyl sulphate
SMALP	styrene maleic acid co-polymer
SPR	surface plasmon resonance
TRIC	temperature related intensity change
TRIS	tris(hydroxymethyl)aminomethane
Tris-NTA	tris-nitrilotriacetic acid

II. List of Symbols

EC_{50}	concentration at which half of the target has bound a ligand
F_{norm}	normalized fluorescence, see MST and TRIC
K_{a}	association constant
K_{d}	dissociation constant
K_{eq}	equilibrium binding constant
K_{i}	dissociation constant measured in a competition experiment
k_{on}	on-rate

k_{off}	off-rate
R	gas constant
T	temperature

III. Introduction

The ability to differentiate between molecular species is essential for a wide array of topics such as solution complexation equilibria, allowing the measurement of equilibrium binding constants and providing information on the stoichiometry of a formed complex. Thermo-optical measurements provide a unique, robust and versatile method to differentiate between different molecular species and are widely used to quantify the strength of the interaction between molecules in solution. The differentiation is based on size, charge and conformational changes, or a combination thereof, that affect a fluorophore-labeled or intrinsically fluorescent molecule, yielding a reliable method applicable to a wide range of potential applications.

Given a general non-covalent, reversible reaction between two molecular species, A and B (Eq. 1), the equilibrium constant, $K_{eq,}$ for the formation of AB is defined by Eq. 2, where the square brackets denote equilibrium concentrations, K_a the association constant (which equals K_{eq}), and K_d the dissociation constant.

$$A + B \rightleftharpoons AB \tag{1}$$

$$K_{eq} = \frac{[AB]}{[A][B]} = K_a = \frac{1}{K_d} \tag{2}$$

Low K_d values (typically in the picomolar to nanomolar range) indicate a high affinity, or tight binding, between the interacting molecules, whereas high K_d values (typically in the millimolar range) indicate a low affinity, or weak binding, between molecules. K_d values can only be determined for non-covalent, reversible interactions at equilibrium. While thermo-optical measurements in principle can also assess covalent binding, these types of interactions will not be addressed here.

Characterizing and quantifying binding affinities are essential to elucidate the molecular mechanisms underlying a myriad of cellular processes and can also be used to design and improve drug-target interactions, antibody-antigen interaction as well as the development of biomolecular sensors.

To determine the equilibrium constant, one needs to quantify [A], [B], and [AB] or know the total concentrations of ligand, $[B]_T = [B] + [AB]$, the total concentration of target $[A]_T = [A] + [AB]$, and the fraction of bound target (or ligand) $x = [AB]/([A] + [AB])$. In thermo-optical measurements, usually the second approach is used and the K_d is determined as follows:

$$x = \frac{[AB]}{[A] + [AB]} = \frac{1}{2[A]_T}\Big([A]_T + [B]_T + K_d - \sqrt{([A]_T + [B]_T + K_d)^2 - 4[A]_T[B]_T}\Big) \quad (3)$$

Using total concentrations has the advantage that they are known a priori so the only parameter that needs to be determined is K_d.

A. Capillary Based Measurements

The most well-known thermo-optical measurement principle is MicroScale Thermophoresis (MST) which is based on the phenomena of thermophoresis, the movement of particles in a temperature gradient, and the general change of fluorescence with temperature. It was first used in binding studies in 2006 and has since become a standard technique to assess molecular interactions.[1–4]

To measure the binding affinity, one of the binding partners is labeled with a fluorophore and will be termed "target" moving forward. The choice which of the two molecules is labeled is arbitrary. The ligand is diluted sequentially in a dilution series and the target is added at a constant concentration. The samples are then loaded into capillaries and placed in a Monolith instrument (NanoTemper Technologies, Figure 1C). For each capillary, the fluorescence is measured before, during and

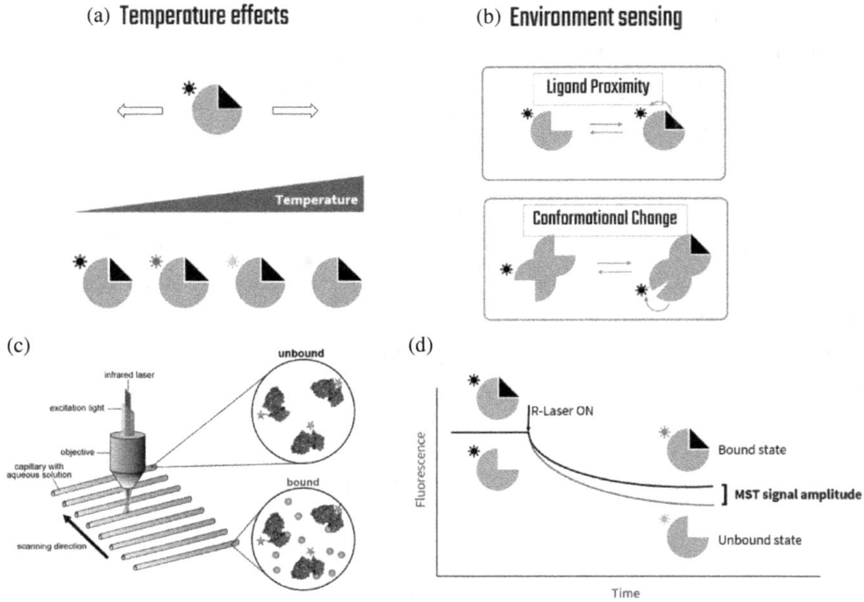

Figure 1. (a) infrared-laser induced temperature effects include the movement of particles in the thermal gradient, thermophoresis, as well as the temperature dependence of the fluorescence intensity. (b) Thermophoresis and the temperature dependence of fluorescence are influenced by the local environment (MST/TRIC), such as ligand proximity, conformational changes, size, and charge and thus sense the environment, which can be used in binding studies. (c) A typical experimental setup using the Monolith instrument showing the capillaries housing samples with constant concentrations of labelled target and increasing concentrations of ligand. The fluorescence at a particular position within the capillary is measured before, during and after local heating with an infrared laser followed by measuring the next capillary. (d) The MST signal amplitude stems from the different temperature response of target in ligand-bound and free states.

sometimes after the heating with an IR-laser and the F_{norm} values (see below) are calculated for a chosen time-point, usually a few seconds after the IR-laser has been turned on (Figure 1D). From these values, the dissociation constant can be determined by substituting Eq. 3 into Eq. 5.

The thermophoretic movement and the temperature sensitivity of the fluorescently labeled target, will differ between free target and complex and thus provide a direct fluorescence read-out of the binding status. The change in fluorescence of a given species can be described by Eq. 4,[3] where c is the concentration of the target molecule, F is the fluorescence intensity of the target molecule per unit concentration, and T stands for the temperature.

$$\frac{\partial}{\partial T}(cF) = c\frac{\partial F}{\partial T} + F\frac{\partial c}{\partial T} \tag{4}$$

The fluorescence will generally decrease with laser-heating as fluorescence shows a strong temperature dependence, meaning $\frac{\partial}{\partial T}(cF) < 0$. The specific amount by which the fluorescence will decrease depends on the environmental sensing through thermophoresis and the environmental sensitivity of the dye (Figures 1A and D). Environment sensing, also called temperature related intensity change (TRIC) of the dye is caused by a change in the local environment of the dye, *e.g.* directly though ligand proximity or through conformational changes of the target molecule (Figure 1B).

As the differences in the temperature-dependence of the fluorescence and thermal movement are minute, even inevitable pipetting inaccuracies would be a source of error. To solve this issue, the fluorescence is normalized to the pre-heating fluorescence intensity, which abolishes most of the concentration dependence of the signal. This normalized fluorescence is called F_{norm} and is dependent on the proportion of ligand-bound target molecules, termed x, as follows:

$$F_{\text{norm}} = x\, F_{\text{norm}}^{(\text{bound})} + (1-x)\, F_{\text{norm}}^{(\text{unbound})}. \tag{5}$$

Stoichiometry measurements and several other specialized assays are described later in this chapter.

B. Multiwell Plate Based Temperature Related Intensity Change Measurements

Advances in understanding how the environment and temperature affects the fluorescence of certain fluorophores have resulted in the development of a measurement technology which is purely based on temperature related intensity change (TRIC), without the need for thermophoresis and therefore without constraints on the sample vessel's dimensions. In instruments of the Dianthus product line (NanoTemper Technologies), the use of environment and temperature sensitive fluorescent dyes allows thermo-optical measurements in multiwell plates, which opens up the possibility for a higher throughput and automation, making it an ideal tool for drug screening. The temperature-dependent fluorescence difference between ligand-bound and free target is small, requiring fast optical heating and utmost precision as well as a well-defined sample to ensure robust experimental data.

Binding and dissociation constants can be determined analogously to MST experiments (see previous paragraph) as both methods yield an optical output that is proportional to the fraction of bound target molecule.

IV. Experimental Guidelines

A. Conducting an MST/TRIC Experiment

As a first step, it is decided which one of the two interacting molecules should be labeled. Factors that govern this decision are typically the availability of both molecules, which labeling strategies/chemistries are possible (*e.g.* accessible His-tag, lysines or cysteines), solubility as the highest obtainable concentration of unlabeled ligand should be 20–50 times above the expected K_d to achieve saturation, and whether there are multiple versions of one of the binding partners to be measured (*e.g.* different protein variants, antibodies, *etc.*).

A single binding affinity measurement requires volumes of about 100–200 μL of typically 10–20 nM of the fluorescent target and 20 μL of the non-fluorescent molecule at its highest concentration.

B. Choosing the Dye and Optimizing Labeling

The success of an MST measurement greatly depends on selecting the right labeling strategy. Consequences of an unfavorable labeling can be insufficient signal-to-noise or data quality or — in the worst case — an inactive target.

For labeling of nucleic acids such as DNA or RNA molecules, as well as peptides, it is best practice to use a label-by-synthesis approach. Here, the fluorescent dye, for example Cyanine 5 (Cy5), can be incorporated during the chemical synthesis of the oligopeptides or oligonucleotides. High-performance liquid chromatography (HPLC) purification after synthesis ensures pure and homogeneously labeled molecules.

When choosing a dye for post-synthesis labeling, it should be noted that superior detection of binding events can be achieved with the so-called "TRIC-sensitive" dyes. These are dyes that show large intensity changes with temperature and are more sensitive to changes in the local chemical microenvironment, which generally can lead to improved signal-to-noise ratios. These second generation dyes are available for several labeling chemistries:

- Lysines and the *N*-terminus of a protein can be labeled using an amine reactive *N*-hydroxysuccinimid (NHS) dye;
- Cysteines of a protein or other free thiols of a molecule can be labeled using a thiol-reactive maleimide dye;
- Polyhistidine tags (*e.g.* His_6- or His_{10}-tags) can be labeled using a dye conjugated to a tris-nitrilotriacetic acid (tris-NTA) moiety;
- Proteins with a SNAP®-tag can be labeled using a dye conjugated to a benzylguanine group.

It is recommended to choose the labeling chemistry that works best with a certain target, but especially for challenging and unstable protein targets, it is a best practice to start with labeling *via* the His-tag using a tris-NTA-conjugated dye if possible. This is a very gentle form of labeling and applicable whenever a protein can be expressed with a polyhistidine-tag. After mixing the dye with the protein in a sub-stoichiometric ratio, no purification is needed, which makes it also possible to label directly in complex samples, for example cell lysates. Limitations of His-tag labeling are incompatibilities with certain buffer substances (such as EDTA) and that very high affinity interactions cannot be measured due to the limited affinity between dye and His-tag. Tris-NTA dyes can be resuspended in a normal PBS buffer and aliquoted and stored at –20 °C.

In case of covalent protein labeling, dye is added to the protein in excess and unreacted dye is removed in a size-exclusion purification step. Afterwards, it is common practice to measure the degree-of-labeling (DOL) of the labeled protein, *e.g.* photometrically. The DOL specifies the average number of dye molecules attached to one protein molecule. Usually, a DOL between 0.5 and 1 is recommended, as a higher DOL indicates that some of the targets are labeled at more than one site.

Common issues for a too low DOL in covalent labeling (NHS or maleimide chemistry) can be that the dye is degraded, not all dye has been properly dissolved or additives unsuitable for the labeling reaction are present in the reaction buffer. In that case, a buffer exchange prior to labeling is necessary and the dye:protein ratio can also be increased. In contrast, if the DOL is too high, the dye:protein ratio can simply be decreased.

Once the labeling reaction has been optimized, one may still fail to obtain a binding curve although an interaction is expected. This can have several reasons. As a first step, it is recommended to test different buffers and buffer additives, including different buffer substances such as PBS, HEPES or TRIS. The choice of buffer can have a large impact on sample quality and especially for interactions, in which unbound and bound state of the target show only a small signal difference, a reduction in noise can make a significant difference. A common pitfall can also be that the buffer lacks necessary additives for the interaction being studied, such as reducing agents or cofactors. In some cases, a change of the labeling strategy can be necessary, particularly when labeling *via* NHS- or maleimide chemistry. In these cases, one needs to consider whether lysine or cysteine residues are located in the ligand binding site which, when labeled, can sterically hinder ligand binding. If there are no alternative labeling possibilities, or if it is unknown which amino acids constitute the binding site, the binding site can be protected with a low-affinity ligand which is present at a high concentration during the labeling step. The ligand is either removed together with unreacted dye in the size-exclusion purification step or dissociates from the target when it is diluted below its K_d concentration.

If an MST/TRIC experiment still is not successful after all these optimization steps, it should be noted that the underlying reason may also be a low protein sample quality. The particular protein sample may have been stored incorrectly, undergone multiple freeze-thaw cycles, or have been subjected to other sub-optimal conditions that may have denatured/unfolded a fraction of the sample. It is therefore a good practice to determine the protein quality beforehand, for example by performing a label-free thermal denaturation assay, dynamic light scattering, *etc.*

C. EC_{50} and K_d

EC_{50} and K_d are sometimes used interchangeably, but actually refer to very different concepts. The EC_{50} is the concentration at which half of the target molecule is bound to a ligand and is a value specific for the experiment. Typically, an EC_{50} is measured when the target concentration is unknown *e.g.* in cell-based receptor binding assays, and its value may change especially when the target concentration is above the K_d.

The K_d is the reciprocal of the equilibrium binding constant K_a or K_{eq}, defined in Eq. 2,

$$K_d = \frac{[A][B]}{[AB]}$$

which is a constant and does not depend on the total target or ligand concentration, but instead relates the reagents' equilibrium concentrations.

It can be shown that the K_d, EC_{50}, and total target concentration $[A]_T$ are related by Eq. 6.

$$EC_{50} = K_d + \frac{1}{2}[A]_T \tag{6}$$

V. Typical Applications

MST/TRIC has been successfully employed to detect interactions between a variety of sample types, such as purified proteins, mammalian and bacterial cell cultures, yeast, fungi, plant extracts and bioliquids. It is often the method of choice since it allows the measurement of both, very high and very low affinities with the same instrument, and also does not require large amounts of samples, as measurements are conducted in small glass capillaries (less than 10 μL in volume) or within tiny wells of a multiwell plate (less than 20 μL in volume). Moreover, it is very flexible regarding assay design and has advantages whenever a target cannot be immobilized on a surface, which is particularly relevant for challenging targets such as multimeric proteins or G-protein-coupled receptors (GPCRs). Often, interactions can be investigated that are hardly accessible by other methods. Since even the smallest changes in the overall charge or hydration energy of a target molecule can lead to a change of the MST signal, it was for example possible to show the binding of nitrate ions to the inner channel of a transmembrane protein.[5] Interactions can be characterized in almost any kind of solvent system, even in complex biological liquids, which guarantees close-to-native conditions and facilitates the study of notoriously difficult-to-handle targets such as transmembrane proteins. Due to this buffer independence, MST/TRIC is routinely used to study interactions in complex environments such as cell lysates, serum, or whole blood.

The MST method is being used extensively to study interactions involving proteins and peptides, nucleic acids (DNA, RNA), many types of small molecules (*e.g.* carbohydrates, fragments) and even ions, but also larger, more complex systems, such as high molecular weight and multidomain complexes,[6] nanoparticles,[7] or liposomes.[8] Apart from simple interaction measurements of binary complexes between two molecules, MST also allows for more complex assays, such as competition experiments, in which the relative affinities of two or more molecules for the same target are measured. Likewise, it allows the investigation of ternary or higher order binding events, *i.e.* interactions that involve three or more binding partners.[9] In addition, stoichiometries between interacting molecules can be determined, *i.e.* the molecular ratios, at which two molecules bind to each other.[10] Another application of MST is monitoring the formation of homo or hetero-oligomers, where a fluorescently labeled monomer is titrated with increasing concentrations of unlabeled monomer. While this approach cannot elucidate the exact form of oligomerization (*i.e.* if dimers or trimers are being formed), it nevertheless allows the measurement of oligomerization constants.[11]

A. Screening for Small Molecule Binding

In contrast to several other fluorescence-based methods, such as anisotropy, fluorescence correlation spectroscopy (FCS) or Förster resonance energy transfer (FRET), the MST/TRIC signal does not depend on changes in molecular size or the relative position of fluorophores toward each other. This major difference extends its applicability beyond only a distinct subset of interactions and makes it a powerful tool for small molecule or fragment-based screening applications. In comparison to other widely applied techniques such as surface plasmon resonance (SPR), it does not suffer from reduced signal-to-noise when measuring low molecular weight fragments. The fluorescence traces in MST can also give information on the presence of aggregates, which gives a fast and direct measure of sample quality. This information has important implications in small-molecule screening campaigns as it can easily help to rule out compounds which induce target aggregation.

The NT.Automated and the Dianthus instruments can be used in hit-identification by either a single dose screening or affinity screening using serial dilutions of the ligand.[12,13] Using the single dose approach, binding is measured at one ligand concentration, typically in duplicate, and the difference of the MST/TRIC signal between a reference sample (no ligand added) and a sample with compound/small molecule is quantified and used for hit selection. In this way, hits can be identified among thousands of ligands. Molecules that are classified as aggregators, auto-fluorescent or quenching compounds can be easily identified and removed from the analyses. In affinity screening campaigns, single dose hits are more thoroughly characterized and typically ranked according to their affinities.

B. Competitive Binding Assays

When measuring the interaction between two molecules with MST/TRIC, one of the molecules is typically labeled with an extrinsic fluorescent dye. Which one of the molecules is labeled can be chosen arbitrarily and should yield identical results, provided that the molecules bind in a 1:1 stoichiometry and labeling does not affect the affinity between the two binding partners. Under certain circumstances, however, labeling may not be possible, for example due to the absence of reactive groups, because the molecules are too small, or the presence of a fluorescent label interferes with binding. In other situations, the intention may be to test many ligands against a single target which cannot be labeled and labeling all the ligands is simply not practical. To circumvent any of the aforementioned limitations, ligand binding to a target can be performed using a competitive approach.

In such a competitive binding assay, the interaction between an unlabeled target and an unlabeled ligand is measured indirectly by the displacement of a labeled competitor molecule from the unlabeled target (Figure 2). In most cases, competitive binding involves competition for the same binding site on the target, but the interaction can also be disrupted by ligand binding to an allosteric binding site. The affinity between target and ligand is usually referred to as K_i instead of K_d to emphasize that it has not been determined in a direct, but in a competitive manner. The K_i can then be calculated by Eq. 7 from the known K_d between

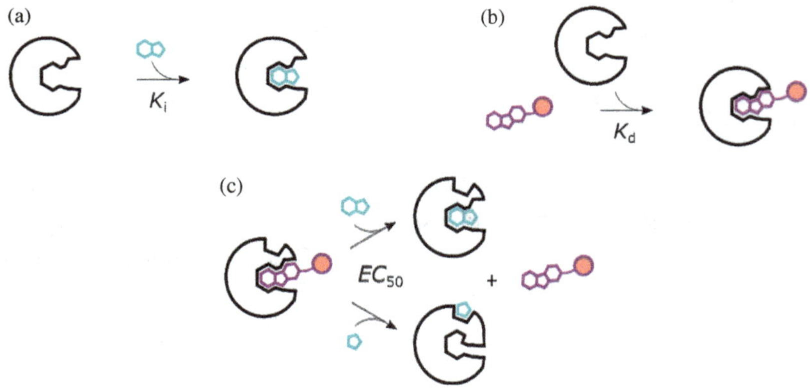

Figure 2. General approach to competition assays. (a) An unlabeled ligand binds to an unlabeled target with unknown dissociation constant K_i. (b) A labeled competitor binds to the same binding site on the target with dissociation constant K_d. (c) Forming a complex between target and competitor and increasing the concentration of unlabeled ligand will displace the competitor from the target, eventually leading to a complex between target and ligand when the concentration of ligand exceeds the EC_{50} concentration. A ligand is called an allosteric inhibitor if it binds to a different binding site while still displacing the competitor from the target.

labeled competitor and target, the EC_{50} of competitor displacement, and the known total concentrations of competitor $[\mathrm{C}]_\mathrm{T}$ and target $[\mathrm{T}]_\mathrm{T}$ in the assay:

$$K_\mathrm{i} = \frac{K_\mathrm{d}}{2-\gamma} \cdot \left(\frac{EC_{50}}{\frac{[\mathrm{T}]_\mathrm{T}}{\gamma} - \frac{K_\mathrm{d}}{2-\gamma} - \frac{[\mathrm{C}]_\mathrm{T}}{2}} - \gamma \right) \tag{7}$$

with

$$\gamma = \frac{[\mathrm{T}]_\mathrm{T} + [\mathrm{C}]_\mathrm{T} + K_\mathrm{d} - \sqrt{\left([\mathrm{T}]_\mathrm{T} + [\mathrm{C}]_\mathrm{T} + K_\mathrm{d}\right)^2 - 4[\mathrm{T}]_\mathrm{T}[\mathrm{C}]_\mathrm{T}}}{2[\mathrm{C}]_\mathrm{T}} \tag{8}$$

Competitive binding-assays are typically conducted in two distinct steps. As a first step, the binding affinity between competitor and target is quantified in a regular binding experiment (Figure 2B). The fluorescent competitor is used at a final concentration around or below the expected K_d which is only limited towards the lower concentration end by the detection limit of the respective instrument, and typically lies between 1 and 20 nM. The concentration is kept constant in all dilutions, while the unlabeled target is titrated up to a concentration of at least 20 times of the expected K_d value. The obtained dose-response can be fitted using the law of mass action to extract the dissociation constant K_d of this interaction (Figure 2A).

As a second step, the ligand of interest is titrated against a complex of target and competitor in a competition experiment (Figure 2C). The target is used at a final concentration which is high enough for complex formation (~1–2 times K_d) and is constant in all dilutions, while the fluorescent competitor is used at the same concentration as in the first experiment. The unlabeled ligand is then titrated into the complex of competitor and target and the displacement of the competitor by the ligand is measured. The resulting data is fitted with the Hill equation and an EC_{50} value is derived for the displacement. Competitive binding is confirmed if the resulting dose-response curve has an opposite direction compared to the binding between target and competitor, *i.e.* if the F_{norm} increases as the population of competitor bound target increases, then the F_{norm} would decrease as the population of ligand bound target increases, and *vice versa* (Figure 3B). Importantly, it is not possible to discriminate between direct competition for the same binding site and an allosteric binding site. In the cases where the ligand binds to another site of the target-competitor complex without fully displacing the competitor (uncompetitive, non-competitive or mixed competitive binding), the dose-response may also change due to the altered chemical environment on the target-competitor complex. However, the F_{norm} value of the bound state will most likely be different to the value of free competitor (Figure 3C). This strategy can also be applied to test if two ligands bind to the same epitope of a target (epitope mapping).

Several examples of competitive MST/TRIC binding assays can be found in the literature. Jerabek-Willemsen *et al.* first applied competitive MST to analyze protein–small molecule interactions.[14] Torres *et al.* later used this assay format to determine binding affinities of heroin hapten-induced antibodies to heroin, its metabolites, and other opioids.[15] Prante *et al.* developed a competitive aptamer-based assay for the detection of the small molecule 25-hydroxyvitamin D in order to maximize the signal-to-noise ratio.[16] Jauset Rubio *et al.* used a competition assay to determine if two DNA aptamers bind to the same or different aptatopes on a protein,[17] while Blech *et al.* applied competitive MST to identify if different therapeutic antibodies bind to the same epitope on interleukin-1β.[18]

C. Emerging Fields

Along with its adaptation in industry and academia, MST has found wide application in numerous research areas that are typically considered challenging.

1. *PROTACs*

MST is routinely used to identify drugs, typically small molecules, that inhibit the function of a certain target. However, many targets are

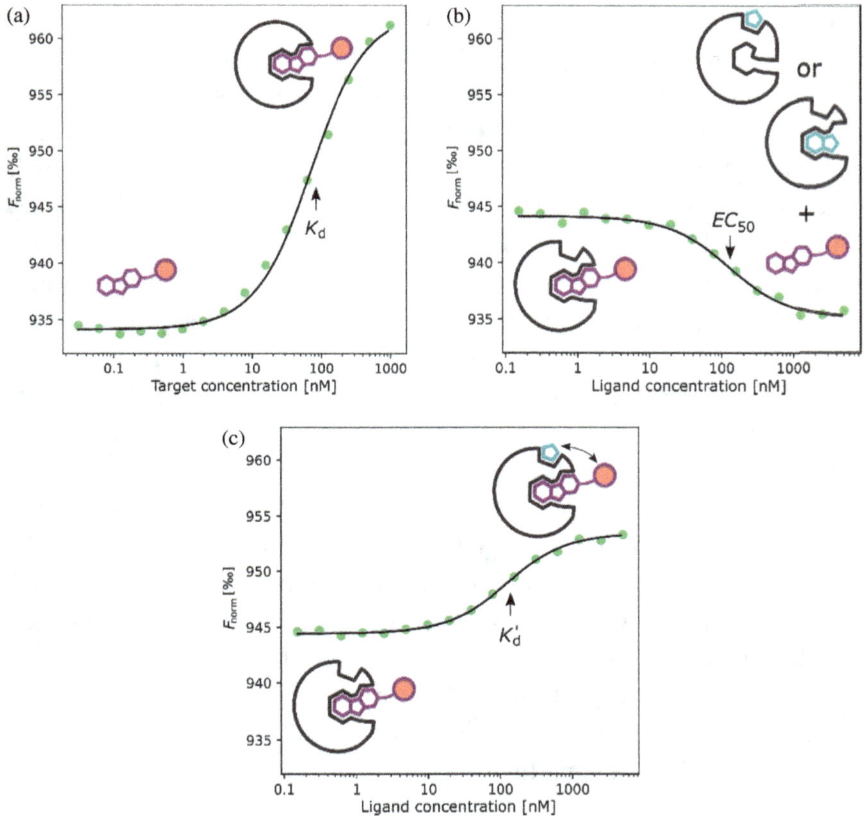

Figure 3. Expected F_{norm} values during a competition assay when (a) the target is titrated against a constant concentration of a fluorescent competitor to determine the dissociation constant K_{d} of the interaction. (b) Target and competitor are mixed at a concentration sufficient to form a complex, roughly at the same concentrations as the K_{d} shown in panel A with an F_{norm} of approximately 945‰, and are then titrated against a series of unlabeled ligand concentrations to obtain the EC_{50}. The displacement can be due to the ligand binding to the same binding site as the competitor, or due to allosteric binding and displacing the competitor indirectly. (c) If ligand binds to a different binding site with dissociation constant K'_{d} without allosterically affecting binding of the competitor, the change in the chemical environment of the fluorophore (indicated by the black arrow) may still cause the dose-response curve to change; however most likely to a different F_{norm} value than the value of free competitor.

considered "undruggable" because so far, not a single ligand has been found. A relatively new approach has therefore been developed to facilitate the degradation of these targets instead of trying to inhibit them. Small molecules such as PROTACs (Proteolysis Targeting Chimeras) and molecular glues can tag a target for degradation by the proteasome, which effectively removes the targeted protein. PROTACs are small, dual-functional molecules that bind to the target protein with one end and to an E3 ligase with the other. The binding to the E3 ligase triggers the ubiquitination of the target protein, marking it for degradation by the proteasome.

To study the working mechanisms of PROTACs, not only binary but also ternary molecular interactions need to be characterized. Surface-based biophysical techniques often fail when it comes to characterizing ternary complexes due to the fact that the second binding partner always has to be present in the solvent at sufficiently high concentrations so that the binary complex is intact when the third binding partner is added. MST as a solution-based method allows one to mix target, PROTAC, and an E3 ligase in defined ratios over a wide affinity range and achieve controlled equilibrium conditions.[19,20] MST can be used to study different modifications to the PROTAC molecule, *i.e.* the E3 ligase recognition motif, the ligand for the target protein or the linker, which can result in varying affinities towards target and ligase.[21]

2. *Gene therapy*

Gene therapy is a growing field of drug discovery research with great promise for the future of pharmaceuticals. To be successful, new gene therapies must effectively and safely deliver nucleic acids to the target cell. This can be achieved by developing high affinity and specific protein binding partners between viral vectors and the target cells.[22]

Another application of MST in the gene therapy field is the analysis of affinities between gene-editing proteins and nucleic acids. Gene editing technologies, which deliver gene therapies *via* targeted *in vivo* genome editing — in particular the CRISPR-Cas9 system — present exciting new possibilities for the treatment of genetic disorders. MST is for example used to make gene editing tools safer by eliminating off-target editing using Cas9 variants.[23]

3. *Neurodegenerative diseases*

A common hallmark of neurodegenerative diseases, like Parkinson's, Alzheimer's, and Huntington's, is the formation of misfolded protein aggregates, which can cause progressive neuronal death, followed by neurological impairment. Understanding the mechanisms behind these processes can help to develop novel therapeutics and early detection methods. MST has become more and more adopted in this field, since it is fast, requires only small amounts of sample and can also provide feedback on aggregation.[24–26] It can differentiate between interactions of monomers, dimers/oligomers and larger fibrils, and the in-solution measurement does not depend on molecular weight changes, which makes it particularly suited to measure binding affinities of small molecules, like aggregation inhibitors or ions to fibrils.

4. *Membrane proteins*

Because of the essential roles they play in cellular processes like signaling and transport, membrane proteins are drug targets with huge potential. Their low expression levels and need to maintain native conformation and solubility after extraction, however, make the characterization of their structure and function — and ultimately, developing drugs that target them — very challenging. MST has made membrane proteins like GPCRs, ion channels, and transporters more accessible for research since its advantages have helped to overcome some of their intrinsic challenges.[5,27–29]

MST is also compatible with the use of nanodiscs and styrene maleic acid lipid particles (SMALPs) for native characterization.[30] To characterize membrane proteins, they are usually removed from their native membrane environment, which can fundamentally change their structure, and therefore their function. Studying membrane proteins in lipid nanodiscs and SMALPs keeps them in a membrane-like environment and without the need of immobilization to a solid matrix.

5. *Plant science*

Performing interaction studies with MST allows for example to characterize the active ingredients of medicinal plants,[31] or to study how plants deal with stress from the environment.[32] MST was also used to study ligand binding properties of periplasmic binding proteins to phosphite and hypophosphite, suggesting the use of phosphite as an alternative phosphorus source in agricultural applications.[33]

6. *Vaccines*

The interactions of a virus with host cell proteins and nucleic acids can be characterized with MST, and also the dimerization and oligomerization of virus proteins themselves, which can be particularly important for antiviral drug development.[34] MST can even be used to determine binding constants using whole virus particles, which can reveal important insights on multivalent binders interacting with the native virus surface.[35]

Furthermore, MST is applied in the discovery of new antiviral drugs, that are used for treating viral infections and developed to block a specific step of the virus life cycle. The drugs bind to a specific site on a virus and through this can interfere with the attachment of the virus to its receptor, replication, or virus assembly.[36,37]

Designing multivalent binders is another promising approach to prevent infection. MST can be used to characterize multivalent binders that inhibit virus infection and validate the optimal design and function of the inhibitor.[38]

VI. Beyond Affinity Measurements

A. Stoichiometry

The stoichiometry is the molecular ratio at which two binding partners interact. While 1:1 interactions are very common, certain interactions, *e.g.* biotin-streptavidin or oxygen-hemoglobin have stoichiometries of 4:1, where each protein binds 4 ligands.

Thermo-optical measurements can be used to determine the stoichiometry of an interaction when the binding sites are equivalent. First, one has to determine the K_d with a standard measurement with the target concentration well below the K_d. In a second step, a binding experiment is performed with the target concentration well above the K_d (*e.g.* 20 times higher). Under these conditions, practically all ligand molecules will be bound to a target molecule at each point in the titration, until no free target molecules are left. This will result in a linear increase of bound target until all the target molecules have their binding sites saturated, and the addition of more ligand will not change the F_{norm} signal. The resulting data can be well approximated with two linear fits, one showing the linear dependence of the F_{norm} to the bound ligand, at lower ligand concentrations, and the other being independent of added ligand indicating saturation, and has a slope close to zero. The intersection of these fits is the approximation of where the target has reached saturation.

The stoichiometry is given by the ratio of the constant target concentration to the corresponding ligand concentration at the point of intersection. An example, the interaction of streptavidin, a homotetramer, with biotin is shown in Figure 4.

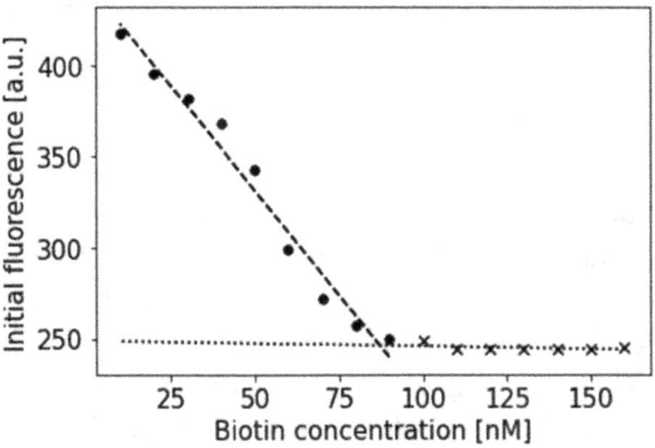

Figure 4. An example of determining the stoichiometry for using fluorescently labeled streptavidin binding to biotin. Raw fluorescence intensities are plotted as a function of ligand concentration and fitted with two independent linear equations (shown by dashed and dotted lines, where the dashed line corresponds to the lower range of ligand concentrations and the dotted line to higher ligand concentrations). The intersection of these two straight lines represents an approximation of the saturation point where the maximum amount of ligand can bind to a known concentration of protein (which is constant at each point). This concentration gives the stoichiometry once it is normalized by the total target concentration. In this case, the linear fits intersect at 87 nM of ligand and the target concentration is 25 nM, resulting in a stoichiometry of 4:1. This corresponds well with streptavidin being a homotetramer, where each subunit has one binding site, and has been shown elsewhere.[39]

B. Van't Hoff Analysis

The dissociation constant (K_d) is arguably the simplest parameter to describe the strength of an interaction between two molecules at equilibrium. Any reversible interaction can be further characterized by its kinetics and thermodynamics. While the binding kinetics, namely the association (k_{on}) and dissociation (k_{off}) rates, report on the rates of complex formation or dissociation, thermodynamic parameters such as the reaction enthalpy ($\Delta_r H$) and entropy ($\Delta_r S$) define the state, in which the Gibbs free energy of a system is the lowest, and provide insight into the molecular mechanisms that drive the interaction.

As a general paradigm in drug development, enthalpy-driven binding is thought to result in more effective drugs with superior physicochemical properties than entropy-driven binding, due to the fact that enthalpy is driven by specific interactions between ligand and target. One approach to derive thermodynamic parameters describing an interaction is from the so-called van't Hoff analysis.

In several cases, the strength of a molecular interaction, K_d, is temperature dependent. Determining dissociation constants at a variety of temperatures enables the calculation of van't Hoff enthalpies by plotting the logarithm of the dissociation constant ($\ln K_d$) *versus* the inverse of the applied temperature (1/T). Van't Hoff plots are typically linear, which is consistent with an invariant binding mechanism across the temperature range studied. The practice of determining van't Hoff enthalpies and entropies from the temperature dependence of the dissociation constant is not inherent to MST, but has also found wide application in other techniques such as SPR, and in many cases thermodynamic constants are similar to those measured by Isothermal Titration Calorimetry (ITC).[40]

At a given temperature T, a system of target and ligand seeks to reach chemical equilibrium. Under this fixed condition, the change in Gibbs free energy ($\Delta_r G$) is related to the K_d by the Gibbs equation (Eq. 9) as follows:

$$\Delta_r G = RT \ln K_d \tag{9}$$

where R is the ideal gas constant and T the temperature in Kelvin. Using the relation between free energy and changes in reaction enthalpy $\Delta_r H$ and entropy $\Delta_r S$, $\Delta_r G$ can be further defined as:

$$\Delta_r G = \Delta_r H - T\,\Delta_r S \tag{10}$$

The $\Delta_r H$ for an interaction is mainly associated with the nature of the interactions at the binding interface. Usually, hydrogen bonds and electrostatic interactions are dominant, although conformational changes can also play an important role. The $\Delta_r S$ for an interaction reflects the degree of order and disorder of the system and is dominated by desolvation of hydrophobic surface areas upon binding.

The variation of the dissociation constant with temperature is then described by the linear form of the van't Hoff equation (Eq. 11).

$$\ln K_d = \frac{\Delta_r H}{RT} - \frac{\Delta_r S}{R} \tag{11}$$

In a van't Hoff plot, ln K_d is plotted *versus* $1/T$, theoretically giving a straight line with slope $\Delta_r H/R$ and y-intercept $\Delta_r S/R$. For endothermic interactions ($\Delta_r H > 0$), ln K_d and thus also K_d decreases as the temperature increases, *i.e.* the molecules bind tighter at higher temperature, while for exothermic interactions ($\Delta_r H < 0$), the opposite is true and molecules bind weaker at higher temperatures. As a side note, this is also one of the reasons, why not every ligand that binds to a protein at room temperature, also leads to a thermal stabilization in a thermal shift assay, as it may simply have already dissociated before the protein reaches a temperature at which it starts to unfold. The maximum temperature at which these experiments can be performed is also limited by the thermal stability of the protein, or biomolecule, and must be taken into consideration.

Limitations of the van't Hoff approach are that the IR-laser induced temperature change can in few cases disturb the chemical equilibrium within the measurement time. This limits the van't Hoff method for MST/TRIC to interactions with slow binding kinetics, which makes it even more important to ensure sufficient equilibration time before starting the measurement. This does not apply for systems, for which the dissociation constant can be determined at constant temperature from either analyzing initial fluorescence changes or the photo-bleaching rate.

Figure 5 shows isothermal binding data obtained from an initial fluorescence analysis of the interaction between a Cy5-labeled 10-mer oligodeoxynucleotide (5' Cy5 GGA CTT CAG G 3') with its unlabeled

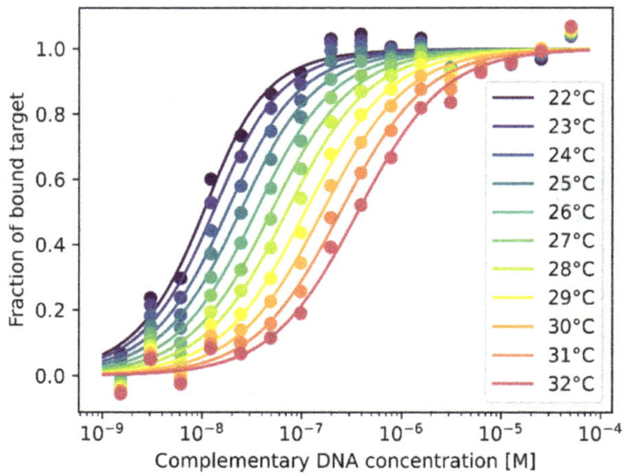

(a)

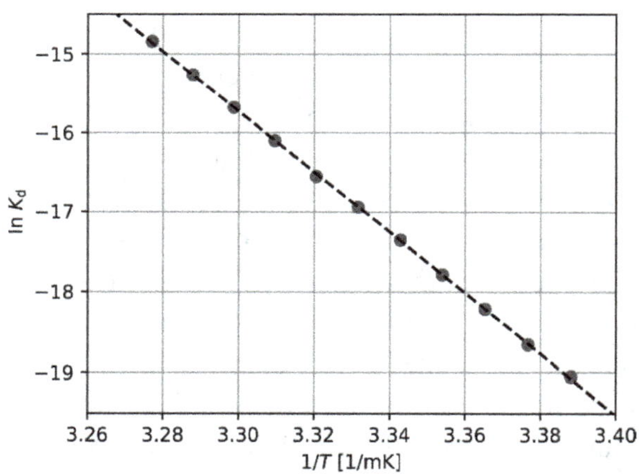

(b)

Figure 5. (a) Dose-response curves of binding affinity measurements between two complementary oligodeoxynucleotides at eleven different temperatures in the range between 22 °C and 32 °C. Binding was measured by analyzing initial fluorescence changes. Dose-response curves shift towards higher concentrations of the complementary strand at higher temperatures, which shows that binding is enthalpically driven. (b) A van't Hoff analysis of the temperature-dependent K_d reveals a linear relationship between ln K_d and $1/T$ (dashed line). The slope of the linear fit can be multiplied by the gas constant R to obtain the enthalpy change $\Delta_r H$ of the interaction, while the intercept needs to be multiplied by R to obtain the standard entropy change $\Delta_r S$ (Eq. 11).

complementary sequence (5' CCT GAA GTC C 3'). Measurements were performed at eleven different temperatures between 22 and 32 °C. As the hybridization reaction between the two strands results in large fluorescence intensity changes, the initial fluorescence signal can be used for K_d determination, resulting in K_d values between 5.3 nM and 359 nM. From a linear van't Hoff analysis (Eq. 11), thermodynamic parameters $\Delta_r H =$ 75.6 kcal mol^{-1} and $\Delta_r S = 218.2$ cal mol^{-1} K^{-1} could be obtained. These values are in good agreement with theoretical calculations.[41]

C. Concentration Determination of Ligand With a Calibration Curve

While MST/TRIC measurements are usually conducted to measure the binding strength of an interaction, the ability of MST/TRIC to differentiate between bound and free species also lends itself well to determining the unknown concentration of a substance when a fluorescently labeled binding partner is available. The first step is to determine the dissociation constant of the interaction. In order to measure in a linear regime, it is imperative to work under stoichiometric conditions, meaning an excess of target. Under these conditions, a calibration curve of known ligand concentrations can be measured. In 2020, Franz *et al.* demonstrated such measurements for the concentration determination of inorganic phosphate, a reaction product of ATPases.[42] In their study, they genetically modified an *E. coli* phosphate binding protein (PBP) to contain a single cysteine, which they used to site-specifically label PBP with a fluorophore. They found a K_d of approximately 0.5 $\propto\mu$M for the binding of phosphate to PBP which can thus act as a phosphate sensor for phosphate concentrations above this K_d. For a linear response of phosphate concentration to the measured signal, they worked under stoichiometric conditions, specifically the phosphate sensor was used at a concentration well above the measured K_d and up to the highest concentration of phosphate that should be determined. In this case, the phosphate sensor was used at 50 μM, which would enable the detection between roughly 1 and up to 50 μM.

Typically, MST measurements are performed with target concentrations in the low nanomolar range and the fluorescent detectors are adjusted to work in this concentration range. Thus, using phosphate sensor concentrations in the micromolar range would saturate the fluorescence detector, assuming all the sensor molecules were labeled. A simple solution is to use only a small sub-population of labeled phosphate sensor molecules, *e.g.* 25 nM labeled and 50 μM unlabeled PBP. In such an experimental setup, only the small proportion of labeled sensor reports binding which is representative for the whole population of sensors.

To determine an unknown concentration, one first records a calibration curve of known phosphate concentrations over the whole range of the assay and into saturation of PBP. In a second step, an unknown concentration can be determined by comparison of the F_{norm} value with the calibration curve, as shown in Figure 6.

D. Measuring Enzyme Kinetics

Most biological processes and increasingly many processes in biotechnology and food technology are mediated by enzymes. Understanding, quantifying, and potentially improving enzyme kinetics is thus of high importance. Measuring enzyme kinetics is generally based on measuring

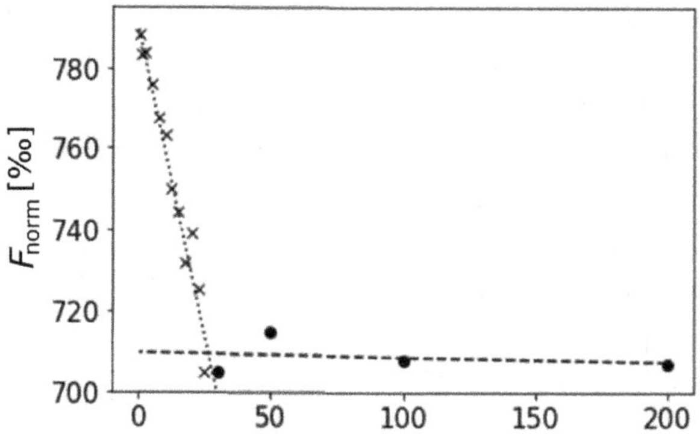

Figure 6. Calibration curve of F_{norm} *versus* phosphate concentration (bound to the PBP) used for the determination of unknown phosphate concentrations. In this example, the total concentration of PBP, a small part of which was fluorescently labeled, was 25 μM. This is approximately the phosphate concentration at which the F_{norm} is independent of additional phosphate, indicating that binding saturation has been achieved (shown by black circles, and fitted to a straight line shown in dashes). The dependence of F_{norm} values on phosphate concentrations below 25 μM are linear (crosses fitted to a straight line, shown by dots) and represents the concentration range in which one can determine unknown phosphate concentrations, under these conditions.

fluorescence or absorbance changes of the substrate(s) or product(s) over time. A limitation to measuring enzyme activity is that the substrate(s) or product(s) may not absorb light or fluoresce. An alternative approach is to use a thermo-optical method, such as MST or TRIC, that can quantify product formation as a function of time. The requirement is that the product of the enzymatic reaction efficiently binds to a fluorescently labeled reporter molecule (biosensor). To determine the concentration of product, such a measurement should be conducted under stoichiometric conditions, where the target (biosensor) concentration should be well above the K_d value (which needs to be determined in a separate experiment) and at a concentration equal to or higher than the maximum concentration of product (ligand) to be measured. This is an extended application of the concentration determination approach described previously.

Franz *et al.* have investigated the enzyme kinetics of an ATPase where the increase in phosphate was measured over time using a fluorescently labeled phosphate binding protein (PBP) which binds to inorganic phosphate.[42] Figure 7 shows the change in F_{norm} over time during the enzymatic hydrolysis of ATP and release of phosphate using such a system. The initial velocity is determined by the linear change of F_{norm} as a function of time, which in this example is between 0 s and 1500 s and has a slope of $-0.0074\ F_{norm}\ s^{-1}$. This slope can be converted into concentration using a calibration curve (see Figure 6). In this case, a decrease of F_{norm} by a factor of 1.27 corresponds to 1 μM change in phosphate concentration. Dividing the slope of $-0.0074\ F_{norm}\ s^{-1}$ by $1.27\ F_{norm}\ \mu M^{-1}$ yields an enzyme activity of $0.0059\ \mu M\ s^{-1}$. These steps can be repeated at different substrate or enzyme concentrations to further characterize the enzymatic reaction.

E. Parallel K_d Measurement and Determination of Ligand Concentration

One of the most important parameters to know when it comes to measuring an accurate K_d is the ligand concentration. However, it is still possible to obtain a $K_{d,}$ with a lower accuracy, using an unknown ligand concentration.

The key lies in the relationship of EC_{50}, K_d, and the total target concentration $[A]_T$, which can be shown to be: $EC_{50} = K_d + \frac{1}{2}[A]_T$ and thus $K_d = EC_{50} - \frac{1}{2}[A]_T$. If two (or more) dose-response curves are

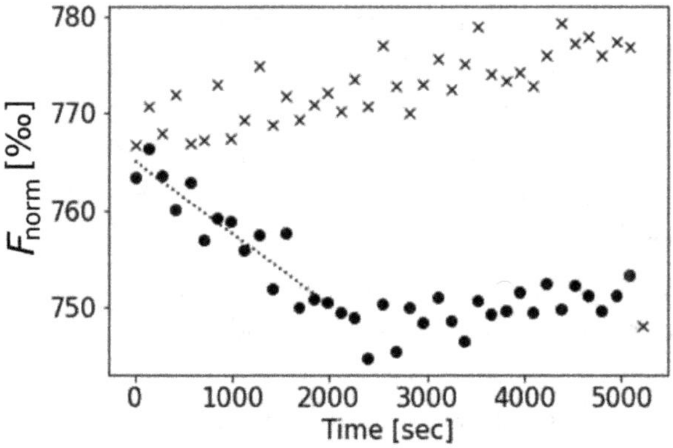

Figure 7. Progress curve showing the change in F_{norm} as a function of time for the enzymatic hydrolysis of ATP (closed circles) and a control reaction with no enzyme added (crosses), where the released phosphate concentration is measured using a fluorescently labeled biosensor. The dotted line corresponds to a linear fit for the first 1500 s of the reaction, with a slope of –0.0074 F_{norm} s^{-1}.

measured with two sufficiently different target concentrations, the K_d can be calculated from the relative position of the two EC_{50} values, although their absolute values are not known. Let's assume we know $[A]_{T1}$, $[A]_{T2}$ and $y = EC_{50,2}/EC_{50,1}$ and that the K_d is identical irrespective of the ligand or target concentration, so:

$$K_{d1} = K_{d2}$$

$$EC_{50,1} - \tfrac{1}{2}[A]_{T,1} = EC_{50,2} - \tfrac{1}{2}[A]_{T,2}$$

$$EC_{50,1} - EC_{50,2} = \tfrac{1}{2}[A]_{T,1} - \tfrac{1}{2}[A]_{T,2}$$

$$EC_{50,1} - y\,EC_{50,1} = \tfrac{1}{2}[A]_{T,1} - \tfrac{1}{2}[A]_{T,2}$$

$$EC_{50,1} = \frac{\left(\tfrac{1}{2}[A]_{T,1} - \tfrac{1}{2}[A]_{T,2}\right)}{1-y}$$

and thus:

$$K_d = \frac{\left(\tfrac{1}{2}[A]_{T,1} - \tfrac{1}{2}[A]_{T,2}\right)}{1-y} - \tfrac{1}{2}[A]_{T,1} \qquad (12)$$

This approach shows, that the K_d can be determined despite unknown ligand concentrations and additionally, the initial ligand concentration can also be determined from knowing the K_d. A global fit of the data points using Eqs. 3 and 5 is an elegant solution and simultaneously allows one to use more than two dose-response curves, increasing the precision of the determined K_d. Figure 8 shows an example of simulated data sets using three target concentrations and Figure 9 an example of real data

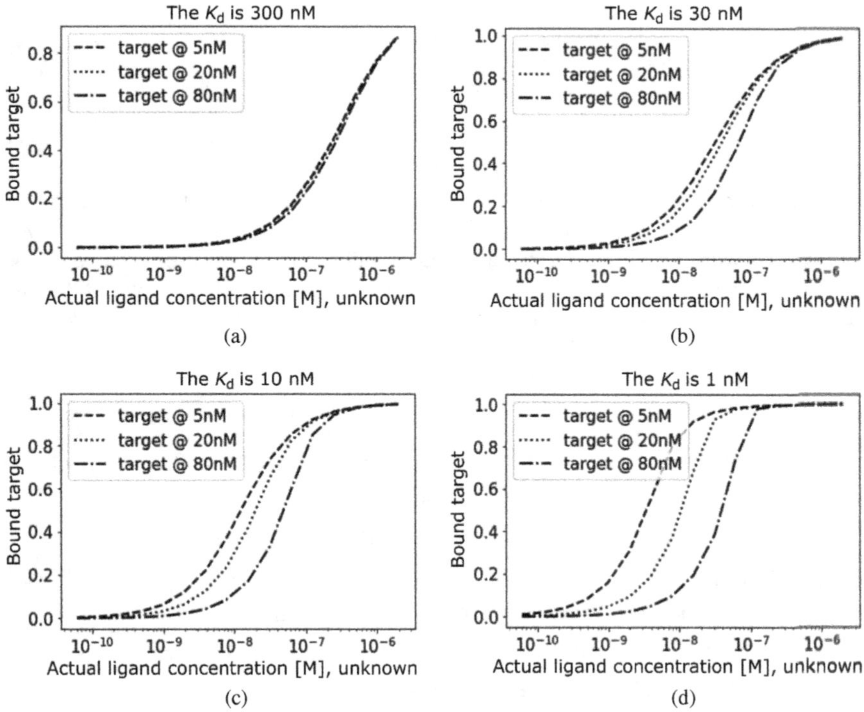

Figure 8. Simulated data showing different scenarios that can be encountered when performing a titration with unknown ligand concentrations and unknown target affinities. Target concentrations need to be chosen spanning the range of the expected K_d. (a) If the target concentrations are too low, one will simply measure the K_d. In such a case, there is no sensitivity for the ligand concentration, but it can be concluded, that the K_d is higher than all the chosen target concentrations. (b, c) Appropriate target concentrations will lead to EC_{50} values dependent on the target concentration, but higher than one-half of the target concentration. In this range, there is sufficient sensitivity to allow the determination of both K_d as well as the ligand concentration. (d) The target concentrations are too high, compared to the K_d, so the EC_{50} is simply one-half of the target concentration. The K_d is lower than the lowest target concentration.

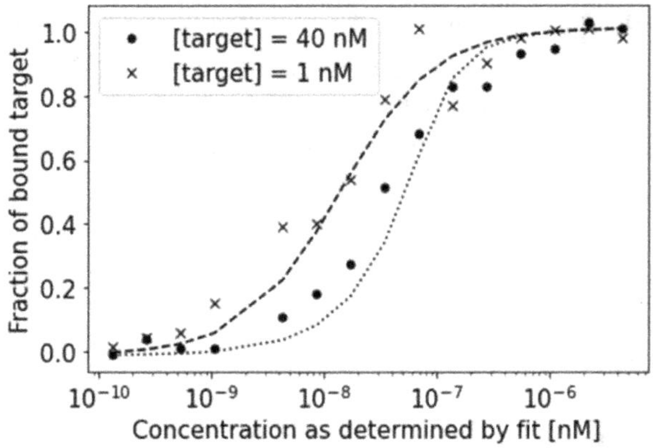

Figure 9. p38α was fluorescently labeled and used at concentrations of 40 nM (closed circles) and 1 nM (crosses). Ligand (SB203580) concentration was spanning 5 μM to 0.15 nM. The fit only used the ligand dilution factor of 2 between data points as well as the target concentration, but not the absolute ligand concentration. As F_{norm} values were deviating slightly, the fitted fraction bound was used instead. Over a wide array of initial fit values, the highest ligand concentration was fitted to roughly 3.5 μM and the K_d to roughly 9 nM. For comparison, the fit using the known ligand concentration of 5 μM found a K_d of 18 and 11 nM for target concentrations of 40 nM and 1 nM, respectively.

recorded with two concentrations of the target protein p38α and its well-characterized inhibitor SB203580.

The need to determine the K_d and the ligand concentration at the same time, can arise in situations such as hybridoma screening (in which many cell lines each produce one type of monoclonal antibody of unknown target affinity and at unknown concentrations) or when determining both concentration and affinity of antibodies in human serum.

An alternative to this approach is to titrate unlabeled target, as described by Lippok *et al.*[43] Depending on the availability of target protein as well as the expected ligand concentration and the expected K_d, the appropriate approach can be chosen.

VII. Practical Tips for Complex Data

A. Ligand Induced Fluorescence Change (LIFC)

As described previously, the MST/TRIC trace requires normalization by the initial fluorescence intensity to compensate for subtle differences in concentration, which arise in large part due to pipetting errors. This implicitly requires that the fluorescence of the target is independent of ligand binding before the sample is locally heated. This requirement is fulfilled in most, albeit not all, cases and where it is found to be significant (in the order of 10–20% or more), it needs to be accounted for. The first step is to ensure that the effect is based on a change of fluorophore brightness and not due to non-specific binding effects of the labeled protein. If denatured sample still shows LIFC after boiling with SDS and DTT, target material was lost, and binding cannot be evaluated. Using low binding reaction vessels and pipette tips, as well as the addition of surfactants, such as polysorbate, can be beneficial in these scenarios to avoid sample loss during handling.

If the denaturation test confirms true LIFC, one can either fit the initial fluorescence intensity directly, as it is proportional to the fraction of bound target molecules, just as the F_{norm} signal, or use an adjusted fitting function, which is more robust with respect to small pipetting errors, see next paragraph.

The F_{norm} values are fitted by taking into consideration LIFC themselves using the following approach (fitting routine has been implemented for software versions MO.Control 2.1, DI.Control 1.0 and DI.ScreeningAnalysis 1.0 and above for both Monolith and Dianthus instruments).

First, the LIFC, meaning the initial fluorescence, is fitted using a sigmoidal fit model. The model is identical to the one used for F_{norm} without LIFC, as both LIFC as well as F_{norm} in the absence of LIFC are proportional to the fraction of bound ligand (Eq. 13),

$$F(x) = F_{\text{b}}x + (1-x)F_{\text{f}} \tag{13}$$

with F being the measured fluorescence and the two parameters F_{b}, the fluorescence of ligand-bound target, F_{f}, the fluorescence of free target. Within the fraction of bound target x, see Eq. 3, lies another fitting parameter, K_{d}.

From this fit, only the ratio, r, of the fluorescence of the bound and of the free state, *i.e.* $r = F_{\text{b}}/F_{\text{f}}$, is used in subsequent steps. This parameter can rely on several points of ideally completely saturated target as well as practically free target and is thus much less sensitive to pipetting errors than the K_{d} or EC_{50}, which can be incorrectly determined by a single outlier in the transition region.

In the next step, the F_{norm} is fitted using Eq. 14,

$$F_{\text{norm}} = \frac{(1 - x)\, F_{\text{norm,f}} + x\, r\, F_{\text{norm,b}}}{1 - x + x\, r} \tag{14}$$

where x again stands for the fraction of bound target (see above) and r for the ratio of the fluorescence intensities of bound target divided by free target, $F_{\text{norm,f}}$ for the F_{norm} value of free target and $F_{\text{norm,b}}$ for the F_{norm} value of bound target. The K_{d} obtained from this is largely independent of pipetting errors and has no bias due to LIFC.

Even when not correcting explicitly for LIFC, the effects on K_{d} are relatively small, as long as the LIFC is small (Figure 10). In a good approximation, the LIFC ratio corresponds to the ratio of actual K_{d} to measured K_{d}, with the deviation increasing slightly if the target concentration is in the range or slightly above the K_{d} value. If the fluorescence intensity of the bound target is only half the fluorescence intensity of the free target, the LIFC ratio is 0.5 and the measured K_{d} will be twice as high as the actual K_{d} (Figure 10). This lies within the error between different

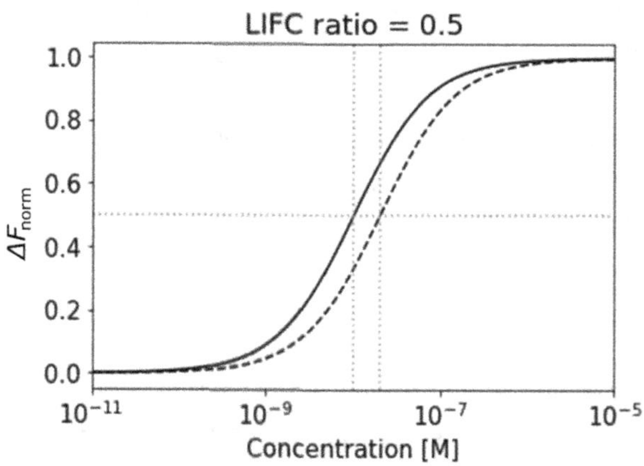

Figure 10. Normalized ΔF_{norm} as a function of ligand concentration (log scale). The dashed line represents simulated data resulting from an LIFC ratio of 0.5. The solid line represents simulated data from an interaction with the same K_{d}, but no LIFC. The EC_{50} differs by a factor of 0.5 as shown by the dotted grey lines.

buffers or methods of K_d determination and a user may choose to not correct for LIFC.

There is one case however, in which LIFC can be problematic, which will be discussed in the next section.

B. Real *Versus* Apparent Biphasic Behavior

In some cases, a target may have more than one binding site for the same ligand and each binding site has a different K_d, resulting in a dose-response curve that is not monotonically increasing or decreasing, but instead shows a peak or a trough. The insight on the existence of biphasic binding, indicating two binding sites, is usually extremely valuable, but at the same time, it is mandatory to do control experiments that independently confirm biphasic binding. The reason for this is, that another phenomenon can lead to dose-response curves that look biphasic: the combination of LIFC and a small shift of the K_d value due to heating of the sample, see Figure 11 for two examples of simulated data sets. With some additional care, this, too, can be fitted and the K_d values of the sample at room temperature as well as during IR-heating can be determined. As the K_d is only defined at equilibrium, the K_d during heating is just an approximation — but the shift can help to identify ligands that both have

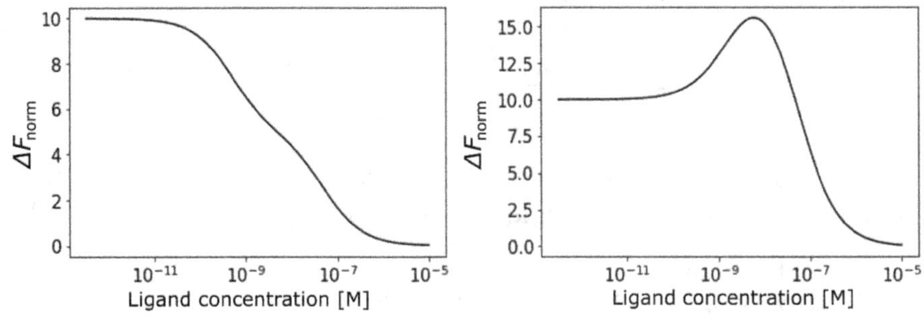

Figure 11. These two sets of simulated data are examples for dose-response binding curves that look biphasic but are actually caused by the combination of LIFC and r of 2 (left) and 0.5 (right) and a K_d change from 1 to 50 nM. Curves similar to those shown on the right can already appear at much smaller r values and smaller K_d changes.

a high off-rate as well as a mainly enthalpic contribution to their free Gibbs energy. So this insight, too, can be valuable.

So why would even a negligible LIFC in combination with a K_d shift, lead to a F_{norm} dose-response curve with a biphasic look to it? Coming back to the definition of F_{norm} (Eq. 15), one can observe that the fluorescence of a sample at a slightly elevated temperature is divided by the fluorescence of a room temperature sample.

$$F_{norm} = \frac{F_{laser\ on}}{F_{laser\ off}} \tag{15}$$

When the K_d changes between these two temperatures, a point of the dilution series with *e.g.* initially 60% bound can turn to be only 40% bound at elevated temperature. This only has an effect if the initial fluorescence is dependent on ligand binding (LIFC).

From Eqs. 15 and 16, one can observe that the fluorescence ratio r of bound and free target at room temperature are contributing to the observed signal. In addition, one also has to consider that there are two different K_d values at different temperatures, which may slightly shift the transition from free to bound target over the dilution series. This equation, can be fitted, with the additional parameter of a second K_d, but as the starting parameters have to be very carefully chosen, this fitting model is not included in NanoTemper Technology Software (see Eq. 16) as of 2021.

$$F_{norm} = \frac{(1 - x_{post})\, F_{norm,f} + x_{post}\, r\, F_{norm,b}}{1 - x_{pre} + x_{pre}\, r} \tag{16}$$

The two bound-fraction functions x_{pre} and x_{post}, are functions of two different K_d values, namely those corresponding to pre- and post-laser heating, respectively.

In many cases, it would be beneficial to have no biphasic behavior. This can often be achieved by changing the labeling strategy, and thus the labeling site, which can completely diminish ligand induced fluorescence changes. This can be achieved by site-specific labeling of protein tags, such as His-tags with a tris-NTA-dye.

VIII. Conclusion

Thermo-optical measurement methods are broadly applicable as they require only small sample volumes, often at low concentrations, are not dependent on size changes of target molecules upon binding exclusively, are performed in solution, and constitute a purely optical method, which makes them contactless, fast, and comparatively cheap. Furthermore, they are highly reproducible as shown in a recent multi-lab study, where the K_d values measured were found to fall within a narrow range of ± 20% around the mean.[4]

Not only can thermo-optical measurements be used for K_d determination but also stoichiometry, thermodynamics, enzyme activity, and much more. The incorporation of ratiometric measurements further increases robustness and applicability to more challenging systems.

IX. References

1. Dhont, J. K. G.; Wiegand, S.; Duhr, S.; Braun, D. Thermodiffusion of Charged Colloids: Single-Particle Diffusion. *Langmuir* **2007**, *23* (4), 1674–1683. https://doi.org/10.1021/la062184m.
2. Duhr, S.; Braun, D. Why Molecules Move along a Temperature Gradient. *Proc. Natl. Acad. Sci. U. S. A.* **2006**, *103* (52), 19678–19682. https://doi.org/10.1073/pnas.0603873103.
3. Baaske, P.; Wienken, C. J.; Reineck, P.; Duhr, S.; Braun, D. Optical Thermophoresis for Quantifying the Buffer Dependence of Aptamer Binding. *Angew. Chem. Int. Ed.* **2010**, *49* (12), 2238–2241. https://doi.org/10.1002/anie.200903998.
4. López-Méndez, B.; Baron, B.; Brautigam, C. A.; Jowitt, T. A.; Knauer, S. H.; Uebel, S.; Williams, M. A.; Sedivy, A.; Abian, O.; Abreu, C.; Adamczyk, M.; Bal, W.; Berger, S.; Buell, A. K.; Carolis, C.; Daviter, T.; Fish, A.; Garcia-Alai, M.; Guenther, C.; Hamacek, J.; Holková, J.; Houser, J.; Johnson, C.; Kelly, S.; Leech, A.; Mas, C.; Matulis, D.; McLaughlin, S. H.; Montserret, R.; Nasreddine, R.; Nehmé, R.; Nguyen, Q.; Ortega-Alarcón, D.; Perez, K.; Pirc, K.; Piszczek, G.; Podobnik, M.; Rodrigo, N.; Rokov-Plavec, J.; Schaefer, S.; Sharpe, T.; Southall, J.; Staunton, D.; Tavares, P.; Vanek, O.; Weyand, M.; Wu, D. Reproducibility and Accuracy of Microscale Thermophoresis in the NanoTemper Monolith: A Multi Laboratory Benchmark Study. *Eur. Biophys. J.* **2021**, *50* (3–4), 411–427. https://doi.org/10.1007/s00249-021-01532-6.

5. Parker, J. L.; Newstead, S. Molecular Basis of Nitrate Uptake by the Plant Nitrate Transporter NRT1.1. *Nature* **2014**, *507* (7490), 68–72. https://doi.org/10.1038/nature13116.
6. Wild, K.; Juaire, K. D.; Soni, K.; Shanmuganathan, V.; Hendricks, A.; Segnitz, B.; Beckmann, R.; Sinning, I. Reconstitution of the Human SRP System and Quantitative and Systematic Analysis of Its Ribosome Interactions. *Nucleic Acids Res.* **2019**, *47* (6), 3184–3196. https://doi.org/10.1093/nar/gky1324.
7. Flayhan, A.; Mertens, H. D. T.; Ural-Blimke, Y.; Martinez Molledo, M.; Svergun, D. I.; Löw, C. Saposin Lipid Nanoparticles: A Highly Versatile and Modular Tool for Membrane Protein Research. *Structure* **2018**, *26* (2), 345–355.e5. https://doi.org/10.1016/j.str.2018.01.007.
8. Starr, M. L.; Sparks, R. P.; Arango, A. S.; Hurst, L. R.; Zhao, Z.; Lihan, M.; Jenkins, J. L.; Tajkhorshid, E.; Fratti, R. A. Phosphatidic Acid Induces Conformational Changes in Sec18 Protomers That Prevent SNARE Priming. *J. Biol. Chem.* **2019**, *294* (9), 3100–3116. https://doi.org/10.1074/jbc.RA118.006552.
9. Arndt, N.; Ross-Kaschitza, D.; Kojukhov, A.; Komar, A. A.; Altmann, M. Properties of the Ternary Complex Formed by Yeast EIF4E, P20 and MRNA. *Sci. Rep.* **2018**, *8* (1), 6707. https://doi.org/10.1038/s41598-018-25273-3.
10. Kazemier, H. G.; Paeschke, K.; Lansdorp, P. M. Guanine Quadruplex Monoclonal Antibody 1H6 Cross-Reacts with Restrained Thymidine-Rich Single Stranded DNA. *Nucleic Acids Res.* **2017**, *45* (10), 5913–5919. https://doi.org/10.1093/nar/gkx245.
11. Ahmed, Z.; Timsah, Z.; Suen, K. M.; Cook, N. P.; Lee, G. R.; Lin, C.-C.; Gagea, M.; Marti, A. A.; Ladbury, J. E. Grb2 Monomer–Dimer Equilibrium Determines Normal *versus* Oncogenic Function. *Nat. Commun.* **2015**, *6* (1), 7354. https://doi.org/10.1038/ncomms8354.
12. Rainard, J. M.; Pandarakalam, G. C.; McElroy, S. P. Using Microscale Thermophoresis to Characterize Hits from High-Throughput Screening: A European Lead Factory Perspective. *SLAS Discovery* **2018**, *23* (3), 225–241. https://doi.org/10.1177/2472555217744728.
13. Schulte, C.; Khayenko, V.; Nordblom, N. F.; Tippel, F.; Peck, V.; Gupta, A. J.; Maric, H. M. High-Throughput Determination of Protein Affinities Using Unmodified Peptide Libraries in Nanomolar Scale. *iScience* **2021**, *24* (1), 101898. https://doi.org/10.1016/j.isci.2020.101898.
14. Jerabek-Willemsen, M.; Wienken, C. J.; Braun, D.; Baaske, P.; Duhr, S. Molecular Interaction Studies Using Microscale Thermophoresis. *ASSAY Drug Dev. Technol.* **2011**, *9* (4), 342–353. https://doi.org/10.1089/adt.2011.0380.
15. Torres, O. B.; Duval, A. J.; Sulima, A.; Antoline, J. F. G.; Jacobson, A. E.; Rice, K. C.; Alving, C. R.; Matyas, G. R. A Rapid Solution-Based Method for Determining the Affinity of Heroin Hapten-Induced Antibodies to Heroin, Its Metabolites, and Other Opioids. *Anal. Bioanal. Chem.* **2018**, *410* (16), 3885–3903. https://doi.org/10.1007/s00216-018-1060-4.

16. Prante, M.; Schüling, T.; Roth, B.; Bremer, K.; Walter, J. Characterization of an Aptamer Directed against 25-Hydroxyvitamin D for the Development of a Competitive Aptamer-Based Assay. *Biosensors* **2019**, *9* (4), 134. https://doi.org/10.3390/bios9040134.
17. Jauset Rubio, M.; Svobodová, M.; Mairal, T.; Schubert, T.; Künne, S.; Mayer, G.; O'Sullivan, C. K. β-Conglutin Dual Aptamers Binding Distinct Aptatopes. *Anal. Bioanal. Chem.* **2016**, *408* (3), 875–884. https://doi.org/10.1007/s00216-015-9179-z.
18. Blech, M.; Peter, D.; Fischer, P.; Bauer, M. M. T.; Hafner, M.; Zeeb, M.; Nar, H. One Target—Two Different Binding Modes: Structural Insights into Gevokizumab and Canakinumab Interactions to Interleukin-1β. *J. Mol. Biol.* **2013**, *425* (1), 94–111. https://doi.org/10.1016/j.jmb.2012.09.021.
19. Ma, D.; Zou, Y.; Chu, Y.; Liu, Z.; Liu, G.; Chu, J.; Li, M.; Wang, J.; Sun, S.; Chang, Z. A Cell-Permeable Peptide-Based PROTAC against the Oncoprotein CREPT Proficiently Inhibits Pancreatic Cancer. *Theranostics* **2020**, *10* (8), 3708–3721. https://doi.org/10.7150/thno.41677.
20. Jin, J.; Wu, Y.; Chen, J.; Shen, Y.; Zhang, L.; Zhang, H.; Chen, L.; Yuan, H.; Chen, H.; Zhang, W.; Luan, X. The Peptide PROTAC Modality: A Novel Strategy for Targeted Protein Ubiquitination. *Theranostics* **2020**, *10* (22), 10141–10153. https://doi.org/10.7150/thno.46985.
21. Haniff, H. S.; Tong, Y.; Liu, X.; Chen, J. L.; Suresh, B. M.; Andrews, R. J.; Peterson, J. M.; O'Leary, C. A.; Benhamou, R. I.; Moss, W. N.; Disney, M. D. Targeting the SARS-CoV-2 RNA Genome with Small Molecule Binders and Ribonuclease Targeting Chimera (RIBOTAC) Degraders. *ACS Cent. Sci.* **2020**, *6* (10), 1713–1721. https://doi.org/10.1021/acscentsci.0c00984.
22. Zhao, Y.; Liu, Z.; Wang, X.; Wu, H.; Zhang, J.; Yang, J.; Zhang, F.; Liu, L.; Long, J.; Lu, P.; Chen, Z. Treatment with Humanized Selective CD19CAR-T Cells Shows Efficacy in Highly Treated B-ALL Patients Who Have Relapsed after Receiving Murine-Based CD19CAR-T Therapies. *Clin. Cancer Res.* **2019**, *25* (18), 5595–5607. https://doi.org/10.1158/1078-0432.CCR-19-0916.
23. Acharya, S.; Mishra, A.; Paul, D.; Ansari, A. H.; Azhar, Mohd.; Kumar, M.; Rauthan, R.; Sharma, N.; Aich, M.; Sinha, D.; Sharma, S.; Jain, S.; Ray, A.; Jain, S.; Ramalingam, S.; Maiti, S.; Chakraborty, D. *Francisella novicida* Cas9 Interrogates Genomic DNA with Very High Specificity and Can Be Used for Mammalian Genome Editing. *Proc. Natl. Acad. Sci. U. S. A.* **2019**, *116* (42), 20959–20968. https://doi.org/10.1073/pnas.1818461116.
24. Fisher, E.; Zhao, Y.; Richardson, R.; Janik, M.; Buell, A. K.; Aigbirhio, F. I.; Tóth, G. Detection and Characterization of Small Molecule Interactions with Fibrillar Protein Aggregates Using Microscale Thermophoresis. *ACS Chem. Neurosci.* **2017**, *8* (9), 2088–2095. https://doi.org/10.1021/acschemneuro.7b00228.
25. Wolff, M.; Mittag, J. J.; Herling, T. W.; Genst, E. D.; Dobson, C. M.; Knowles, T. P. J.; Braun, D.; Buell, A. K. Quantitative Thermophoretic Study of Disease-Related Protein Aggregates. *Sci. Rep.* **2016**, *6* (1), 22829. https://doi.org/10.1038/srep22829.

26. Agerschou, E. D.; Flagmeier, P.; Saridaki, T.; Galvagnion, C.; Komnig, D.; Heid, L.; Prasad, V.; Shaykhalishahi, H.; Willbold, D.; Dobson, C. M.; Voigt, A.; Falkenburger, B.; Hoyer, W.; Buell, A. K. An Engineered Monomer Binding-Protein for α-Synuclein Efficiently Inhibits the Proliferation of Amyloid Fibrils. *eLife* **2019**, *8*, e46112. https://doi.org/10.7554/eLife.46112.
27. Ural-Blimke, Y.; Flayhan, A.; Strauss, J.; Rantos, V.; Bartels, K.; Nielsen, R.; Pardon, E.; Steyaert, J.; Kosinski, J.; Quistgaard, E. M.; Löw, C. Structure of Prototypic Peptide Transporter DtpA from *E. Coli* in Complex with Valganciclovir Provides Insights into Drug Binding of Human PepT1. *J. Am. Chem. Soc.* **2019**, *141* (6), 2404–2412. https://doi.org/10.1021/jacs.8b11343.
28. Yoshida, K.; Nagatoishi, S.; Kuroda, D.; Suzuki, N.; Murata, T.; Tsumoto, K. Phospholipid Membrane Fluidity Alters Ligand Binding Activity of a G Protein-Coupled Receptor by Shifting the Conformational Equilibrium. *Biochemistry* **2019**, *58* (6), 504–508. https://doi.org/10.1021/acs.biochem.8b01194.
29. Bada Juarez, J. F.; Muñoz-García, J. C.; Inácio dos Reis, R.; Henry, A.; McMillan, D.; Kriek, M.; Wood, M.; Vandenplas, C.; Sands, Z.; Castro, L.; Taylor, R.; Watts, A. Detergent-Free Extraction of a Functional Low-Expressing GPCR from a Human Cell Line. *Biochim. Biophys. Acta, Biomembr.* **2020**, *1862* (3), 183152. https://doi.org/10.1016/j.bbamem.2019.183152.
30. Bernhard, M.; Laube, B. Thermophoretic Analysis of Ligand-Specific Conformational States of the Inhibitory Glycine Receptor Embedded in Copolymer Nanodiscs. *Sci. Rep.* **2020**, *10* (1), 16569. https://doi.org/10.1038/s41598-020-73157-2.
31. Wu, S.; Yang, Y.; Liu, B.; Xie, Z.; Xiong, W.; Hao, P.; Xiao, W.; Sun, Y.; Ai, Z.; Wu, H. A Novel Anti-Platelet Aggregation Target of Chinensinaphthol Methyl Ether and Neojusticin B Obtained from *Rostellularia procumbens* (L.) Nees. *J. Enzyme Inhib. Med. Chem.* **2019**, *34* (1), 999–1009. https://doi.org/10.1080/14756366.2019.1609468.
32. Wang, P.; Zhao, Y.; Li, Z.; Hsu, C.-C.; Liu, X.; Fu, L.; Hou, Y.-J.; Du, Y.; Xie, S.; Zhang, C.; Gao, J.; Cao, M.; Huang, X.; Zhu, Y.; Tang, K.; Wang, X.; Tao, W. A.; Xiong, Y.; Zhu, J.-K. Reciprocal Regulation of the TOR Kinase and ABA Receptor Balances Plant Growth and Stress Response. *Mol. Cell* **2018**, *69* (1), 100–112.e6. https://doi.org/10.1016/j.molcel.2017.12.002.
33. Bisson, C.; Adams, N. B. P.; Stevenson, B.; Brindley, A. A.; Polyviou, D.; Bibby, T. S.; Baker, P. J.; Hunter, C. N.; Hitchcock, A. The Molecular Basis of Phosphite and Hypophosphite Recognition by ABC-Transporters. *Nat. Commun.* **2017**, *8* (1), 1746. https://doi.org/10.1038/s41467-017-01226-8.
34. Zeng, Z.; Deng, F.; Shi, K.; Ye, G.; Wang, G.; Fang, L.; Xiao, S.; Fu, Z.; Peng, G. Dimerization of Coronavirus Nsp9 with Diverse Modes Enhances Its Nucleic Acid Binding Affinity. *J. Virol.* **2018**, *92* (17), e00692-18. https://doi.org/10.1128/JVI.00692-18.
35. Lauster, D.; Glanz, M.; Bardua, M.; Ludwig, K.; Hellmund, M.; Hoffmann, U.; Hamann, A.; Böttcher, C.; Haag, R.; Hackenberger, C. P. R.; Herrmann, A.

Multivalent Peptide-Nanoparticle Conjugates for Influenza-Virus Inhibition. *Angew. Chem. Int. Ed.* **2017**, *56* (21), 5931–5936. https://doi.org/10.1002/anie.201702005.

36. Ascher, D. B.; Wielens, J.; Nero, T. L.; Doughty, L.; Morton, C. J.; Parker, M. W. Potent Hepatitis C Inhibitors Bind Directly to NS5A and Reduce Its Affinity for RNA. *Sci. Rep.* **2015**, *4* (1), 4765. https://doi.org/10.1038/srep04765.
37. Sierra, N.; Folio, C.; Robert, X.; Long, M.; Guillon, C.; Álvarez, G. Looking for Novel Capsid Protein Multimerization Inhibitors of Feline Immunodeficiency Virus. *Pharmaceuticals* **2018**, *11* (3), 67. https://doi.org/10.3390/ph11030067.
38. Lauster, D.; Klenk, S.; Ludwig, K.; Nojoumi, S.; Behren, S.; Adam, L.; Stadtmüller, M.; Saenger, S.; Zimmler, S.; Hönzke, K.; Yao, L.; Hoffmann, U.; Bardua, M.; Hamann, A.; Witzenrath, M.; Sander, L. E.; Wolff, T.; Hocke, A. C.; Hippenstiel, S.; De Carlo, S.; Neudecker, J.; Osterrieder, K.; Budisa, N.; Netz, R. R.; Böttcher, C.; Liese, S.; Herrmann, A.; Hackenberger, C. P. R. Phage Capsid Nanoparticles with Defined Ligand Arrangement Block Influenza Virus Entry. *Nat. Nanotechnol.* **2020**, *15* (5), 373–379. https://doi.org/10.1038/s41565-020-0660-2.
39. Hendrickson, W. A.; Pahler, A.; Smith, J. L.; Satow, Y.; Merritt, E. A.; Phizackerley, R. P. Crystal Structure of Core Streptavidin Determined from Multiwavelength Anomalous Diffraction of Synchrotron Radiation. *Proc. Natl. Acad. Sci. U. S. A.* **1989**, *86* (7), 2190–2194. https://doi.org/10.1073/pnas.86.7.2190.
40. Day, Y. S. N.; Baird, C. L.; Rich, R. L.; Myszka, D. G. Direct Comparison of Binding Equilibrium, Thermodynamic, and Rate Constants Determined by Surface- and Solution-Based Biophysical Methods. *Protein Sci.* **2002**, *11* (5), 1017–1025. https://doi.org/10.1110/ps.4330102.
41. Breslauer, K. J.; Frank, R.; Blocker, H.; Marky, L. A. Predicting DNA Duplex Stability from the Base Sequence. *Proc. Natl. Acad. Sci. U. S. A.* **1986**, *83* (11), 3746–3750. https://doi.org/10.1073/pnas.83.11.3746.
42. Franz, P.; Gassl, V.; Topf, A.; Eckelmann, L.; Iorga, B.; Tsiavaliaris, G. A. Thermophoresis–Based Biosensor for Real–Time Detection of Inorganic Phosphate during Enzymatic Reactions. *Biosens. Bioelectron.* **2020**, *169*, 112616. https://doi.org/10.1016/j.bios.2020.112616.
43. Lippok, S.; Seidel, S. A. I.; Duhr, S.; Uhland, K.; Holthoff, H.-P.; Jenne, D.; Braun, D. Direct Detection of Antibody Concentration and Affinity in Human Serum Using Microscale Thermophoresis. *Anal. Chem.* **2012**, *84* (8), 3523–3530. https://doi.org/10.1021/ac202923j.

https://doi.org/10.1142/9789811281099_bmatter

Index

www.ingramcontent.com/pod-product-compliance
Lightning Source LLC
LaVergne TN
LVHW010334220926
848484LV00004BB/897